ISR 56

Interdisciplinary Systems Research
Interdisziplinäre Systemforschung

Derek W. Bunn

The Synthesis of Forecasting Models in Decision Analysis

1978 Birkhäuser Verlag, Basel und Stuttgart

CIP-Kurztitelaufnahme der Deutschen Bibliothek

Bunn, Derek W.:
The synthesis of forecasting models in decision analysis. – 1. Aufl. – Basel, Stuttgart: Birkhäuser, 1978.
(Interdisciplinary systems research; 56)
ISBN 3-7643-1009-X

ISBN 3-7643-1009-X

PREFACE

A specifically decision-making approach is taken in this work towards the utilization of forecasting models. Based largely upon an analysis of the implications of the Coherence Principle of rational belief and action, which is fundamental to current Bayesian Decision Analysis methods, a somewhat radical forecasting methodology is advocated. It is concluded that a sensible utilization of forecasting models for decision analysis requires a formal synthesis of the set of applicable models. This methodology is evidently in contrast to much of the effort of forecasting theorists who, for a long time, have looked to the scientific method for guidelines on how to select the best predictor and, by implication, to disregard the others. A great deal has been written on forecasting methods and decision analysis, but very little attention has been given to the implications which each embodies for the other and, in particular, upon the sensible utilization of the evidence from forecasting models within decision analysis.

Part I, therefore, discusses the basic principles for decision analysis and forecasting and argues the rationality of a synthesis of forecasts. Part II examines various methods to this end and Part III provides several case study examples to indicate their feasibility. The computational algorithms are listed in the Appendix.

Much of the initial work on this project was undertaken at the London Graduate School of Business Studies and later at the International Institute for Applied Systems Analysis in Austria. This work was completed in the Department of Decision Sciences at the University of Southern California. The author wishes to acknowledge the support of all these institutions and also the competence of the Word Processing Center at the latter in the final preparation of the typescript.

CONTENTS

PART I: FOUNDATIONS

The basic principles and methodology of Bayesian Decision Analysis and Forecasting are briefly reviewed and the fundamental position of Coherence is emphasized. It is recognized that in the general context of policy formulation, the decision maker must effect a coherent synthesis of all of the available evidence including, particularly, forecasting models.

1.1 An Outline of the Decision Analysis Approach

The "Decision Analysis" approach to formulating a decision problem under uncertainty is now conventionally understood to imply the application of the set of methods advocated by the Subjectivist Bayesian theories of rational choice and inference. With a considerable debt to the earlier work of von Neumann and Morgenstern (1948) and Savage (1954), the principal expositions of this methodology were to be found in Luce and Raiffa (1957) and Raiffa and Schlaifer (1961). Since the mid-sixties, the decision analysis approach has become perhaps the most widely acceptable way of dealing with a great variety of problems.

These problems currently range from strategic policy formulation (cf. Bunn and Thomas, 1978) to tactical organizational management (cf. Moore, et al., 1976). Reported case studies of implementation span a diversity from, for example, the siting of airports (Keeney, 1973), or nuclear power plants (Keeney and Nair, 1976), river pollution (Ostrom and Gros, 1975), forestry (Bell, 1975) to Space research (Judd et al., 1974) and medical care (Aitchison and Kay, 1976). Decision Analysis is now taught in most universities and business schools with textbooks ranging from those aimed at the practicing manager, such as Thomas (1972) or Lindley (1971), to LaValle (1970) and de Groot (1970) for university courses in statistics.

The basic principle of rationality underlying Decision Analysis is that of Coherence of Preferences. This means that any opinion or preference held by the decision maker should not contradict all the other opinions or preference held. Lindley (1971) justifies this principle quite simply:

> The coherence is easily defended. Suppose we had an incoherent person who said E is less likely than F, F is less likely than G and, instead of concluding that E is less likely than G, he concludes that G is less likely than E. Consider again a situation in which he is to receive a prize if E occurs, and otherwise not. Then, keeping the prize fixed throughout the discussion, he would prefer to base his receipt of the prize on F, rather than E, and indeed would pay you a sum of money (or part of the prize) if you would substitute F for E. Accept the money and replace E by F. Now the argument may be repeated and you can receive more money by replacing F by G. Having gained two sums of money, you can now offer to replace G by E and the person would want to accept since he regards G as less likely than E. So a third sum of money passes and we are back to the situation we started from where the prize depended upon E. The incoherent person is back where he started except that he has given you some money. The cycle may be repeated if he holds to his uncertainty relations, and the incoherent person is a perpetual money-making machine. This shows that incoherence is untenable.

Coherence is clearly, therefore, a <u>necessary</u> condition for rational belief but whether it is <u>sufficient</u> in itself as a principle for rationality is a controversial issue in current philosophy. This will be discussed further in the next section.

Decision Analysis works by prescribing a coherent ranking of all the decision options on the basis of relatively few stated preferences by the decision maker. This has been achieved essentially by means of adapting the Bayesian use of subjective probability in, for example, Jeffreys (1948) to the Expected Utility criterion of von Neumann and Morgenstern (1948).

It is perhaps easiest to demonstrate this by means of a simple illustration. Suppose that the decision maker must select the "best" investment (a^*) from the set of options (A). It is assumed that each option is evaluated only with respect to the financial measure of Net Present Value. The outcome of each option therefore represents a point on the real line (V). The decision maker will be uncertain on the outcome for each option but it is presumed that he can identify the points V_L and V_U which he feels sure represent lower and upper bounds on V for all the prospective outcomes in (A).

Furthermore, it is also assumed that the decision maker can articulate his uncertainty for each option (a_i), where all elements of (A) have been numbered through the index i, in terms of a subjective probability distribution $F_i(V)$, expressed in the cumulative form. Practical aspects of assessing subjective probabilities are discussed in Bunn (1975), Bunn and Thomas (1976), Stael von Holstein (1970) and Winkler (1967).

A convenient method, recently advocated by Barclay and Peterson (1973), requires the decision maker to state the tertiles of $F_i(V)$. These are the two points within the range (V_L, V_U) which split V into three equiprobable intervals. Thus, the points $F_i(V) = 0; = 1/3; = 2/3; = 1$ are defined and intermediate points inferred by interpolating a smooth curve drawn through them. A more detailed discussion of estimating subjective probabilities is given in the Appendix since the use of subjective probabilities is an important theme in this work.

A utility function is also assessed upon V in order to reflect the decision maker's attitude towards risk. A decision maker should always prefer that option for which the expected utility is highest. The establishment of a utility function is an attempt to quantify the other main factor influencing decision making under uncertainty which is that of value. The method generally used is based upon that of von Neumann and Morgenstern (1948). The value of an uncertain prospect is derived by finding a Certainty Equivalent, i.e., a prospect with no uncertainty associated, which also obeys the Expectation operation as a mathematical property. If $U(V_L) = 0$ and the other extreme outcome, $U(V_U) = 1$ by construction, then $V_{.5}$ for which $U(V_{.5}) = 0.5$ is obtained as the decision maker's certainty equivalent for the uncertain prospect of V_L or V_U with equal probabilities. This follows from the expected utility principle:

$$\begin{aligned} U(V_{.5}) &= 0.5\ U(V_L) + 0.5\ U(V_U) \\ &= 0.5 \end{aligned} \tag{1.1}$$

Similarly

$$\begin{aligned} U(V_{.25}) &= 0.5\ U(V_L) + 0.5\ U(V_{.5}) \\ &= 0.25 \end{aligned} \tag{1.2}$$

and

$$\begin{aligned} U(V_{.75}) &= 0.5\ U(V_U) + 0.5\ U(V_{.5}) \\ &= 0.75 \end{aligned} \tag{1.3}$$

From these points, the smooth function U(V) can again be obtained by interpolation. The most desirable option a* can now be prescribed as

$$\underset{a_i \in A}{\text{Max}} \left[\int_{V_L}^{V_U} U(V)\ d\ F_i\ (V) \right] \tag{1.4}$$

Clearly, there are weak coherence assumptions in interpolating the probability and utility functions on (V) and a strong coherence assumption in using these functions to evaluate the expected utility of (a_i) and hence deduce the preference ranking of (A).

The value of the decision analysis approach lies therefore in the framework it provides to allow the decision maker to arrive at a demonstrably coherent action on the basis of several relatively easily explicated preferences. In the above example, rather than attempting the difficult conceptual task of a full ranking of (A) on a purely intuitive basis, the decision maker only stated his preferences for the few carefully chosen and apparently more easily conceptualized situations needed to construct the utility and probability measures. The appropriately coherent ranking of (A) is then deduced.

The power of the Decision Analysis approach to provide a coherent structure for what otherwise would be an entirely intuitive process and to break up a complex decision task into a set of smaller and more manageable assessments becomes even more evident when the above Subjective Expected Utility principle is embedded in a decision tree representation and combined with Bayesian methods of probability revision. Decision trees are particularly useful in providing a full visual representation of the ramifications of the problem and also an algorithm for its solution. Brown, et al. (1974) provides an easy introduction to decision tree analysis. The Bayesian methods of probability revision facilitate the incorporation of quantitative information with subjective assessments. They, in a sense, provide coherent learning models to allow the modification of subjective "prior" assessments, through the impact of data, to "posterior" assessments. Lindley (1971) provides an introduction to the Bayesian principles and LaValle (1970) gives a more thorough treatment.

It should be noted, however, that these methods of Decision Analysis do not "solve" the problem for the decision maker. He must still provide the preference judgments upon which the prior subjective probability distributions and the utility functions are imputed. All that Decision Analysis does is to provide a coherent structure which allows a complex problem to be divided up into separate assessment tasks. This is, of course, of great value in helping to focus attention on critical aspects and in understanding their interrelationships. However, the extent to which the approach is normative in giving the decision maker directives for action is only at the level of coherence. The Decision Analysis approach does not, for example, indicate what a decision maker's basic attitude to risk *should* be. This is inherent in the preferences he articulates in the original construction of the utility curve. Decision Analysis then goes on to deduce the coherent implications of this risk attitude in ranking the decision options. A correctly formulated Decision Analysis only embodies, therefore, assumptions of a logical character. This absence of behavioral assumptions is a strength from the theoretical point of view but may be a weakness from the pragmatic perspective of the practicing decision maker.

1.2 Further Implications of Coherence

The practicing decision maker can sometimes find himself rather disillusioned with the methods of Decision Analysis. When confronted with the utility assessment task, for example, he will frequently be unsure of what the appropriate attitude to risk should be for a rational person in his position. He may have instinctive responses to the assessment tasks, but he recognizes that a thinking man would be wary of these, particularly as it should be recalled that the particular preference judgments required are usually somewhat artificially contrived within the decision situation. Similarly, in the probability assessment tasks, the decision maker may find it difficult to formulate a precise opinion on the uncertainties. There may be a considerable amount of disparate, diverse and even perhaps conflicting evidence available such that the decision maker feels that he needs more help in formulating a rational synthesis.

A very cynical view of Decision Analysis might imply that it apparently only serves to restructure the problem and has little to offer on the crucial aspects of formulating the basic preferences. What Decision Analysis gains by separating out various aspects, it distorts by the artificiality of the contrived assessment tasks.

These criticisms appear to hold a great deal of validity, but it will be demonstrated in the following sections how the approach of Decision Analysis is, in fact, being extended in an attempt to alleviate these practical difficulties.

The essential value of the Decision Analysis approach, it will be recalled, is in providing an explicit structure to facilitate a coherent analysis of the problem. The methodology provides checks on the coherence of the explicated preferences and decision-tree representation of the problem. These consistency checks, as they are often called, ensure that the explicit preferences are not subject to the irrationality of a "Dutch book" as exemplified by the reference to Lindley's perpetual money-making machine in the previous section.

Thus, Decision Analysis is rational only as far as coherence in the preferences and structure as articulated by the decision maker. This coherence of revealed preferences is a very limited sanction of rationality. It is necessary, of course, but far from sufficient.

Tversky (1974) recognized this at a practical level when he suggested that decision analysts should not satisfy themselves with the usual consistency checks. A coherent set of preferences can still be irrational. For example, a set of preferences for outcomes in a coin tossing experiment could be mutually coherent and yet still based on the "gambler's fallacy" of assuming the probability of, say, heads on a particular toss increases with the number of consecutive tails that proceed it. Thus, Tversky argued that it is not sufficient to reveal an internally coherent set of preferences and suggests that these responses should be consistent with all the other beliefs held by the individual. In the coin tossing example, the responses are incompatible with the belief that an unbiased coin has no memory.

Tversky is here recognizing at a practical level one of the fundamental requirements for the existence of a subjective probability measure, namely that of the total coherence of all the beliefs held by an individual. Savage (1954) found that in order to derive a fine probability measure purely on the basis of preference judgments, he had to ensure the individual's coherence over a "large world" and not just the "small world" of the revealed preferences in question. Furthermore, Good (1962) emphasizes that in assigning subjective probabilities over even an apparently straightforward set of events, this requirement of total coherence implies that the task is implicitly one of measuring an unmeasurable set.

Thus, an individual can never be sure that his coherent set of revealed preferences are totally coherent with his complete set of evidence and beliefs. Fishburn (1964) is making this point when he states that despite the Behaviorist assumption that an individual always reveals his true preferences, he is frequently not sure of the objectives and bases upon which to formulate his belief. The implication is that with a fuller ratiocination, his revealed preferences may well be different.

Hence, the necessity of pursuing this unattainable requirement of total coherence helps to explain the difficulties of the practicing decision maker in formulating his preferences. That total coherence is a necessary condition for rational choice has been demonstrated as inescapable. According to Rescher (1973), although it has not yet been proven to be sufficient, total coherence is also a more sensible theory of rationality than any alternative, particularly in the context of decision making. Whether it is indeed a sufficient condition to guarantee rationality is in practice of no consequence given that it is unattainable in any case. The important point, however, is that rationality requires the pursuit of total coherence.

Decision Analysis methodology must therefore extend itself in order to encourage a more explicitly coherent formulation of the preference judgments which are the decision maker's inputs into the usual style Decision Analysis.

The specific concern of this research is the coherent formalization of the subjective probability inputs in the Decision Analysis model. It should be mentioned in passing, however, that a considerable amount of work in this spirit is currently being undertaken on the utility side through developments in the practical techniques of multiattributed utility measurement. Keeney and Raiffa (1976) is a good reference on this.

The explicit formulation of subjective probability, on the other hand, has received relatively scant attention, although the topic of hierarchical (sometimes called "cascaded") inference is beginning to attract increasing investigation in the psychological literature. In 1974, the journal _Organizational Behavior and Human Performance_ devoted a special issue to the topic and Peterson (1973) has recently published a monograph. In hierarchical inference, some attempt is made to break up the assessment task into separate assessments conditioned upon the structure of the hypotheses and the overall inductive model. This is particularly relevant to estimating the probability of rare events, as for example in connection with problems of reliability and failure in the nuclear power industry (cf. Selvidge, 1973). A somewhat different application of a "thinking algorithm" in subjective probability assessment has recently been developed by Bunn and Thomas

(1975) in the context of competitive bidding.

In general, however, it would appear that explicit consideration of the formulation of subjective probability, particularly in the area of policy analysis where it involves establishing a probability measure on the set of conceivable futures, must inevitably relate to the utilization of forecasts. Thus, the investigation of methods for the explicit formalization of the inductive hypotheses which underwrite a subjective probability estimate in the standard decision analysis effectively involves the incorporation of the theory and methodology of forecasting into the Decision Analysis paradigm. Decision Analysis would therefore appear to have much to gain from the adoption of a forecasting approach to estimating the probability inputs. More significantly, perhaps, is the considerable gain to the conventional theory and methodology of forecasting from the accession to a Decision Analysis orientation. Sensible procedures of policy analysis will, moreover, necessitate such a consolidation of the disciplines of decision theory and forecasting.

1.3 Principles of Forecasting Methodology

A general set of principles for rational forecasting have still, it seems, yet to be formulated. Burns (1974), for example, states:

> Despite the increasing demands from management for explicit and systematic forecasts and the growth in the application of statistical techniques, forecasting remains substantially an informal practice. In part this is due to the lack of a general adequate theory of forecasting and, hence, each problem often appears "unique" and requiring special attention.

Granger and Newbold (1973) are more categorical when they state:

> No well-defined theory of forecasting, with a general information set or arbitrary cost function, exists.

Hence, there seems to be an emphasis upon the uniqueness of each problem and a tendency to regard predictive model-building as having a largely artistic element. Imagination and creative thinking must clearly play an important part in the conceptualization of the various possible future scenarios, but this does not preclude the use of rational analysis in the construction and evaluation of the associated predictive models. Most decision problems are unique and moreover require a considerably broad perspective in conceiving all of the ramifications, yet Decision Analysis, through the Coherence and Subjective Expected Utility principles, provides a general scheme of rational evaluation.

Unfortunately, however, rather than looking for underlying general principles, methodological guidelines for practical forecasting have tended to be developed through ad hoc empirical evidence. The various heuristics proposed for adaptive exponential smoothing are typical examples (cf. Brown, 1963; Trigg and Leach, 1967; Roberts and Whybark, 1974).

Jenkins (1974), de Finetti (1970) and Good (1965) have all condemned what is often referred to as "adhockery" in statistical methods. Good and de Finetti have provided firm theoretical bases for rational inference but without the translation into practical methodology. Box and Jenkins (1970) have proposed probably the most outstanding practical methodology for a large class of time-series predictors (i.e., the ARIMA family), but this is still a restricted field within the general scope of forecasting models. ARIMA predictors, moreover, require considerable judgment in selection and the accumulation of quite a large time-series of data (50 or more, in general). Many of the techniques of forecasting have been developed to a very high level of sophistication but in relative isolation as separate topics in their own right. The methods of spectral analysis on time series data or multiple regression models on economic data are examples of this. Separate techniques have apparently been developed without general principles of consolidation.

Thus, the practitioner is today faced with a wide and disparate set of predictive techniques with which to approach his forecasting problem. These are well documented in the many texts on business forecasting [cf. Benton (1972); Bolt (1971); Robinson (1971), Wolf (1966)] or the ICI monographs of the 1960's on technological forecasting [cf. Martino (1972); Wills (1971) or the OECD review of 1972] or any of the specialist management areas such as corporate planning, marketing or production planning.

Many writers of forecasting texts have looked to the scientific method to provide a rational basis. Theil (1970) and Mincer and Zarnowitz (1969), for example, begin their treatises with attempts to establish a scientific forecasting theory. At a practical level the scientific "taxonomy and selective treatment" theme appears common. The distinctive features of forecasting situations and the particular characteristics of possible forecasting models are identified with the aim of being able to prescribe the "right" model for particular situations. Adams (1973), Chambers et al. (1971) and Makridakis and Wheelright (1973) pursue this theme and provide the appropriate forecasting model/situation allocation matrices of varying degrees of comprehensiveness.

A certain amount of false esteem can apparently be reflected upon a method by contriving to establish it with a scientific predicate. There has been a great temptation, therefore, for promoters of the newer and less well-developed fields of inquiry, such as forecasting, to rush into the scientific method without questioning the appropriate assumptions. An analysis of the implications of pursuing the scientific method in the forecasting context will be continued in the following paragraphs purely to demonstrate the full extent of the difficulties it brings when used without a full awareness of its basic assumptions.

One dilemma which confronts this pursuit of the scientific method and the consequent selective approach to forecasting is the problem of <u>deciding</u> upon a criterion of forecast evaluation. Even restricting consideration to measures of forecast accuracy, the practitioner is confronted with many possible evaluators. Theil (1970), Mincer and Zarnowitz (1969) and Gadd and Wold (1967) have all proposed differing measures such as mean deviation or mean squared error. With these measures ranking a set of predictive models differently, Tan (1971), for instance, found it a difficult task to select out the best approach in his consideration of the prediction of UK electricity demand. Theil's $\underset{\sim}{U}$ statistic is often used in evaluating econometric models, but can easily be shown to be inconsistent with a MSE criterion.

The problem can easily be resolved by a Bayesian. Attention could be focused on defining a utility function on forecast errors such that the best method to be selected will be that which maximizes expected utility. But, unfortunately, the scientific method precludes the incorporation of individual value judgments. Scientific methodology is concerned with isolating the model of highest verisimilitude and not with personal preferences. The decision to reject all but the most "truthful" model should be based upon rational principles rather than a personal loss function. This isolation of forecasting from decision making is quite commonly advocated. Thus, Granger and Newbold (1975) state, with some apparent conviction:

> One thing that we would certainly argue is that the forecasting task can be completely separated from that of the decision maker.

If we pursue this quest for a scientific method further, principles are clearly now required to enable us to select out the best criterion of forecast evaluation. Evidently, there is going to be an infinite regress and the problem enters the realm of inductive logic. Indeed, there is a fundamental philosophical problem relating to the formulation of rational inference in the light of empirical evidence, namely the old Problem of Induction.

1.4 The Problem of Induction

I shall use the wide interpretation of induction as referring to the generalization of a hypothesis outside the direct field of experience and this will therefore encompass all nontrivial types of forecasting. Katz (1962) summarizes the fundamental problem thus:

> We make one prediction rather than another because we inductively extrapolate past uniformities . . . We are tempted to say that we judge the reliability of this policy by looking at the record. But, as Hume points out, this option is not open because we cannot legitimately justify the general policy of appealing to experience by appealing to experience. Since it is this very type of appeal to the record whose justificability is in question, justificatory arguments based on what has been experienced merely go in a circle and beg the very question at issue.

In other words, empirical grounds are insufficient for validating a predictive method. It is recognized, however, that empiricism can perform the weaker function of vindicating a forecast in the sense that a method can be demonstrated to have performed better than another. But we cannot then go on to conclude that such performance justifies the assertion that it will perform better in the future. Bayesians, of course, have the facility to argue that such evidence can revise the relative probability of one method outperforming the others. This appears to be the key to the problem but again it involves a slightly different perspective to that of orthodox scientific method.

It seems rather odd that the spirit of the scientific method still persists in seeking to identify the "true" hypothesis. If the first quarter of this century has taught us anything about the nature of physical science (and physics is after all the science all others try to emulate), it is, through the work of people such as Heisenburg and Einstein, the futility of searching for absolute measures in anything but closed logical systems. Yet Popper (1959 and 1963), for example, persists with a two-valued logic, attempting to label a hypothesis as true or false, and rejecting systems of many-valued logic (which involve the assignment of probabilities to the hypotheses) on the grounds that a probability measure will be misleading. Using a logical interpretation along the lines of Carnap (1952), he argues that models with the highest probabilities will be those with the most general implications and hence those which say least. This is clearly the case with most interpretations of probability that refer to the inherent truthfulness of the models but not necessarily for those based upon the relative performance of the models. It is in fact the performance based interpretations which are the more relevant to decision orientated prediction. One such formulation is given in Bunn (1975) and further developed in Section 2.3. Furthermore, it is also true that in Popper's scientific code of conjectures and refutations, the most general hypotheses are necessarily the least falsifiable and would therefore be the last to be rejected from the "true" set of models.

The crucial point of departure of the scientific method with policy-orientated prediction can now be seen. The ultimate aim of the scientific method is to provide rules for the isolation of falsifiable hypotheses and not the selection of the most "truthful" after all. Thus, the scientific epistemologists are inclined to justify induction under a set of rules for choosing a hypothesis whose experimentation is likely to contribute most to scientific knowledge rather than one which is expected to be the best predictor for use in decision analysis.

1.5 Forecasting within Decision Analysis

A decision-making perspective on forecasting implies a different underlying resolve. When confronted with a decision problem, the decision analyst wishes to utilize forecasts in order to prescribe the optimum decision. In this instance, he is not in the scientific role of postulating and testing descriptive models in order to gain a greater understanding of the world. Rather, his approach is Pragmatic and Instrumentalist.

It is in fact relatively easy to justify induction on Instrumentalist grounds. Thus, Reichenbach (1949) states:

> If predictive methods cannot supply a knowledge of future, they are nevertheless sufficient to justify action.

It has already been pointed out that modern Bayesian inference has the facility to put a probability measure on a set of inductive models. This would appear more sensible than the arbitrary discrimination in, for example, the frequentist procedure of model falsification. In this method, experiments conducted at a significance level of α% imply that α% of the models labelled as falsified should in fact have remained in the true set. Furthermore, a two-valued logic would not permit such a falsified model to return for consideration in the true set. Bayesian model discrimination schemes have been used by Giesel (1974), Zellner (1971) and Wiginton (1975). A probability measure is given to each forecasting model to represent its relative forecasting ability with respect to the others in the set being considered, and this measure is revised with experimental evidence according to Bayesian principles. The scheme does not embody a mechanism to label all but one of the set as falsified. The analyst is free to deal as he wishes with the models on the basis of their relative posterior probabilities.

A hint was given earlier that the Bayesian decision theoretic approach precludes entering this controversy of model discrimination through the direct possibility of applying the expected utility criterion to model performance. The argument was protracted, however, in order to expose the fundamental error in seeking to adopt the scientific method. Thus, it is now evident that such an expected utility approch would still be contaminated by the scientific selective theme. The fundamental decision problem is not one of model selection but of model utilization with the function of the forecasts being to provide informative subjective probability inputs into the decision analysis.

Modern decision theory places considerable importance upon the accumulation of information and the articulation of as informative a prior density as possible. The greater the precision of the subjective probability distributions, the more effective the decision analysis can be in evaluating the options. Moreover, posterior precision will generally be proportional to the information content. Thus, the underlying resolve of the decision analyst is to utilize the optimal amount of information. Furthermore, the Total Coherence requirement of a subjective probability measure, referred to previously, dictates that any subjective probability used in a decision analysis should reflect consistency with the totality of evidence and beliefs held by the individual at that time.

Thus, uncertainty resolution in decision and policy analysis requires that the subjective probability inputs should be judgments based upon a rational synthesis of all the forecasts and opinions that it makes economic sense to accumulate. A basic task must therefore be to develop an operationally feasible methodology for effecting a sensible formal synthesis of forecasts.

1.6 A Formalization of Coherent Forecasting within Decision Analysis

The total coherence of all beliefs held by a decision maker is a necessary condition to justify rational beliefs. The previous sections have indicated two important implications of this principle. Firstly, if the subjective probability inputs to a decision analysis are to be an explicit formalization of as much as possible of the total evidence and opinions available, then they must embody a clear consideration of all the forecasting models that it makes economic sense to formulate. Secondly, this consideration will reflect a synthesis of the models and not a selection of the "best" on its own.

Incidentally, the optimum number of forecasting models to consider can be evaluated accordingly to the standard Bayesian approach to the value of information as in Raiffa and Schlaifer (1961). Since a forecasting model is performing the sole function of increasing the information content of the subjective probability input to the decision analysis, there is, in principle, no difference between this and the standard case study examples of evaluating prospective market research, etc.

In the following sections we will assume that the set of forecasting models under consideration (M) does in fact represent the optimal. Similarly (X) represents the optimal amount of data. If the operation of model m_i on datum x_i is denoted by the pair (m_i, x_i), then inference based upon full utilization of models and data will be formulated on the product set (M x X). Evidently, some models will be inappropriate to certain data and thus certain subsets of (M x X) will infer nothing.

The distinction between (M) and (X) appears clear enough when we think of forecasting models such as multiple regression, 1st order exponential smoothing, Gompertz curve, etc., and data as time series or random samples. When, however, consideration is turned to the subjective subset of (X), the well-known philosophical problem of the interrelation of theories and facts could actually become a practical difficulty. Both psychologists (cf. Neisser, 1967) and philosophers (cf. Popper, 1959), argue that all observations are theory laden; that is, that no "raw" data is absolutely free of interpretation. In most instances this is a somewhat pedantic point of view and an unhelpful one if the argument is pursued to the Skeptic or even Solipsist position of invalidating all empirical knowledge. In any case, the decision orientation imposes a Pragmatic position where the emphasis is not upon the truthfulness of some information but upon whether it is the best available. Thus, the distinction is postulated as a potentially useful conceptual device insofar as it can establish an intuitively acceptable procedure of rational inference.

To pursue this abstract formalization further, total coherence implies that explicit account should not only be taken of the set of predictive models (M) and all the relevant data (X) but also particular attention must be given to the evidence (A) that the decision maker might have the relative appropriateness of each model in the given situation. Clearly (A) would represent the information content of the probability measures, discussed in 1.5, used in Bayesian model discrimination schemes. Thus, the decision maker's aspirations to total coherence require him to make an explicit synthesis of as much of (M x X x A) as methodology will allow.

PART II: METHODOLOGY

It follows from the coherence requirements of rational decision analysis that any method for the synthesis of forecasting models should follow a Bayesian procedure. Two classes of methodology are discussed in this part. The first is based upon deriving the optimum parameters for a linear synthesis with the assumption of a minimum variance objective function. A second class of methods uses a subjective probability measure defined on the forecasting models to represent their relative performance. Both these classes of methods are evaluated by analysis and simulation experiments.

2.1 The Synthesis of Forecasting Models

There is nothing intuitively or behaviorally novel in the pursuit of the synthetic approach. At an informal and unstructured level, decision makers inevitably form a consensus opinion upon the variety of different forecast estimates with which they are confronted. The motivation for a more formal procedure is also apparent. Thus, we frequently see on the sports pages of newspapers a consensus football results forecast and in the political discussion, a pooling of the opinion poll estimates.

Through this intuitively felt need by decision makers for a synthesis of forecasts, much of the early methods developed along *ad hoc* guidelines. The absence of a more fundamental methodology can, of course, be ascribed to the way in which the analysis of forecasting principles was confused by the scientific method.

In formulating a synthesis of forecasting models, consideration is generally limited to the class of linear unbiased predictors. Thus, if the variable being predicted is denoted by (y) and $f_i(y|m_i)$ is the unbiased predictor of (y) based upon model (m_i), then the synthesized predictor, $f_s(y|M)$, is given by

$$f_s(y|M) = \sum_{i=1}^{n} k_i f_i(y|M_i) \tag{2.1}$$

if the (M) contains n models indexed through i, and where the n x 1 vector, k, denotes the set of weights used in this linear combination. Usually, only the mean of the synthesized predictor is required and as an unbiased point estimate, $E[f_s(y|M)]$, it is given by

$$E(F_s(y|M)) = \sum_{i=1}^{n} k_i \; E[f_i(y|m_i)] \tag{2.2}$$

The only constraint on k is that

$$\sum_{i=1}^{n} k_i = 1 \tag{2.3}$$

There have been two important lines of analysis in estimating the vector, k, of linear weights:

(i) A Minimum Variance Synthesis

(ii) A Subjective Probability Synthesis

These two approaches will be discussed in sections 2.2 and 2.3 respectively.

2.2 A Minimum Variance Synthesis of Forecasts

The optimal linear weights, $\underset{\sim}{k}*$ under a minimum variance criterion for two forecasting models with error variances σ_1^2, and σ_2^2 and correlation coefficient ρ are:

$$k_1* = (\sigma_2^2 - \rho\ \sigma_1\ \sigma_2)/(\sigma_1^2 + \sigma_2^2 - 2\ \rho\ \sigma_1\ \sigma_2) \tag{2.4}$$

$$k_2* = (\sigma_1^2 - \rho\ \sigma_1\ \sigma_2)/(\sigma_1^2 + \sigma_2^2 - 2\ \rho\ \sigma_1\ \sigma_2) \tag{2.5}$$

Proof:

From equations 2.1 and 2.3,

$$f_s(y|M) = k_1\ f_1(y|m_1) + (1 - k_1)\ f_2(y|m_2) \tag{2.6}$$

Thus,

$$\sigma_s^2 = k_1^2 \sigma_1^2 + (1 - k_1)^2 \sigma_2^2 + 2\rho k_1(1 - k_1)\sigma_1 \sigma_2 \qquad (2.7)$$

By simple differential calculus,

$$d/dk_1(\sigma_s^2) = 2k_1 \sigma_1^2 - 2(1 - k_1) \sigma_2^2 + 2(1 - 2k_1)\rho \sigma_1 \sigma_2 \qquad (2.8)$$

Equating this to zero for the minimum gives $\underset{\sim}{k} = \underset{\sim}{k}^*$ as equations 2.4 and 2.5.

Using these optimum weights, $\underset{\sim}{k}^*$, the variance of the synthesized predictor, σ_s^2, will never be greater than either σ_1^2 or σ_2^2 for any value of the correlation coefficient taken from its admissable range, i.e. $-1 \leq \rho \leq +1$.

Proof:

Using k* from equations 2.4 and 2.5, σ_s^2 is evaluated as

$$\sigma_s^2 = [\sigma_1^2 \sigma_2^2 (1 - \rho^2)]/(\sigma_1^2 + \sigma_2^2 - 2\rho\sigma_1 \sigma_2) \qquad (2.9)$$

Because of symmetry we can assume $\sigma_1 \leq \sigma_2$.

Subtract σ_s^2 from σ_1^2 to give, after rearranging the formula,

$$\sigma_1^2 - \sigma_s^2 = \sigma_1^2 (\sigma_1 - \rho\sigma_2)^2/[(\sigma_1 - \rho\sigma_2)^2 + \sigma_2^2(1 - \rho^2)] \qquad (2.10)$$

which is clearly ≥ 0.

Corollary:

The synthesis will give no reduction in the minimum variance predictor $f_1(y|m_1)$ whenever, following equation 2.10,

$$0 = \sigma_1(\sigma_1 - \rho\sigma_2^2) \qquad (2.11)$$

This will hold if

Case 1: $\sigma_1 = 0$ (2.12)

i.e. one of the models is a perfect predictor, or

Case 2: $\sigma_1 = \sigma_2$ and $\rho = 1$ (2.13)

i.e. the two models are statistically identical, or

Case 3: $\rho = \sigma_1/\sigma_2$ (2.14)

This last case is perhaps the most surprising in showing that for even the nontrivial instance of a pair of different positive forecast variances, there will always be one particular value of the correlation coefficient for which a synthesis yields no improvement upon the best individual predictor.

However, the optimal minimum variance predictor will generally show quite a significant reduction in the variance of the best individual predictor, as equation 2.10 demonstrates. In the extreme cases, as $|\rho|$ approaches unity, σ_s^2 tends to zero. Equation 2.9 also shows that in the case of two independent, identically distributed forecasting models, the error variance can be reduced to a half by the simple linear synthesis of taking the average of the two models. As a typical example, Figure 2.1 displays the synthesized variance as a function of ρ for the case of $\sigma_1 = 1$; $\sigma_2 = 2$.

The significance of synthesizing models does in fact increase considerably with the number of models considered. The above results for $n = 2$ can be generalized for any n. Adapting a result by a Halperin (1961) from the theory of the Minimum Variance Unbiased Estimate (MVLUE) to the forecasting context, the optimal n x 1 vector, $\underset{\sim}{k}^*$, is given by

$$\underset{\sim}{k}^* = \underset{\sim}{S}^{-1} \underset{\sim}{E}/(\underset{\sim}{E}' \underset{\sim}{S}^{-1} \underset{\sim}{E}) \tag{2.15}$$

where $\underset{\sim}{S}$ is the n x n covariance matrix of forecast errors and $\underset{\sim}{E}$ is the n x 1 unit vector. Furthermore, if $\underset{\sim}{F}$ is the n x 1 vector of posterior means of $f_i(y|m_i)$, then the minimum variance synthesized forecast f_s^* is given by

$$f_s^* = (\underset{\sim}{E}'\underset{\sim}{S}^{-1}\underset{\sim}{F})/(\underset{\sim}{E}'\underset{\sim}{S}^{-1}\underset{\sim}{E}) \tag{2.16}$$

Clearly, for computational simplicity, if $\underset{\sim}{C}$ is the cofactor matrix of $\underset{\sim}{S}$, then the vector of unnormalized weights $\underset{\sim}{W}$ is given by

$$\underset{\sim}{W} = \underset{\sim}{C}\underset{\sim}{E} \tag{2.17}$$

and thus

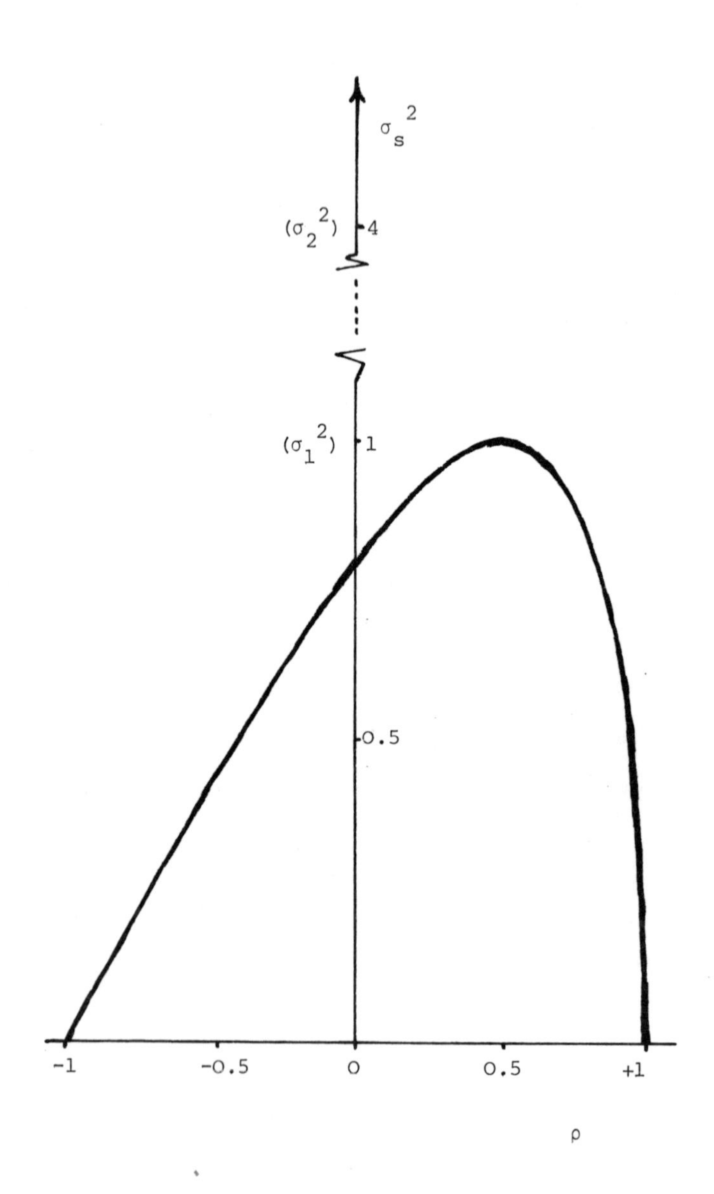

FIGURE 2.1 THE SYNTHESISED VARIANCE σ_s^2, OF TWO PREDICTORS, $\sigma_1^2 = 1$; $\sigma_2^2 = 4$, AS A FUNCTION OF ρ.

$$f_s^* = \underset{\sim}{W}'\underset{\sim}{F}/tr(\underset{\sim\sim}{WE}'). \qquad (2.18)$$

One very considerable problem with the use of equation 2.16 is the estimation of $\underset{\sim}{S}$. A substantial set of data is required which must also be stationary over time. Because of these practical difficulties, Bates and Granger (1969) and Newbold and Granger (1974) considered a collection of heuristic methods for effecting the linear combination. Their suggestions were based on two principles, viz. that most weight should be given to the forecast which has performed best in the recent past and that the linear weights should adapt to the possibility of non-stationarity over time between the performance of the individual predictors.

Thus, their approach is much akin to the heuristic methods commonly employed to find the best predictor in the short-term, adaptive forecasting context, referred to in Section 1.3. Moreover, similar criticisms also appear applicable. Although the large number of case studies investigated clearly demonstrated the efficiency of combining forecasts, a certain amount of doubt has been expressed on the generalizability of the results (c.f. the discussion on Newbold and Granger, 1974). This is always a problem with a purely empirical approach to forecasting along scientific selective lines for the reasons discussed in Section 1.4. Furthermore, in these studies, it was found difficult to discriminate between the various synthetic heuristics which were suggested.

The argument contained in Part One demonstrated that in the decision context, forecasting methods must be subject to the rationality principles of Decision Analysis. A particular implication of the total coherence axiom is that a synthetic procedure for combining forecasts must be justifiable within Bayesian methodology.

Bayesian Estimation of S

The Bayesian approach is to describe the error covariance matrix by a subjective probability distribution which embodies all the relevant information that the decision maker might have on $\underset{\sim}{S}$ and in such a way that this prior distribution can be readily updated with the impact of new data.

The appropriate natural conjugate probability density function with which to encode this subjective information on the multinormal covariance matrix $\underset{\sim}{S}$ will be, following the terminology of Lavalle (1970), the nth order Inverted Wishart with parameters v (scalar) and $\underset{\sim}{\Psi}$ (n x n).

$$f(\underset{\sim}{S}|\underset{\sim}{\Psi},n) = W_n(v)^{-1} |\underset{\sim}{\Psi}|^{\frac{1}{2}(v+n-1)} \; |\underset{\sim}{S}^{-1}|^{\frac{1}{2}(v+2n)} \exp(-\tfrac{1}{2}v\,tr\,(\underset{\sim}{S}^{-1}\underset{\sim}{\Psi}) \quad (2.19)$$

with

$$W_n(v) = (2/v)^{n(v+n-1)/2}\pi^{n(n-1)/4} \prod_{i=1}^{n} \Gamma[\tfrac{1}{2}(v+n-i)] \quad (2.20)$$

defined for $\underset{\sim}{S}$ positive definite and symmetric and $v > 0$

$$\text{mean } (\underset{\sim}{S}) = \underset{\sim}{\Psi}v/(v - 2) \quad (2.21)$$

$$\text{mode } (\underset{\sim}{S}) = \underset{\sim}{\Psi}v/(v + 2n) \quad (2.22)$$

This distribution is evidently an n-dimensional generalization of the inverse gamma density function.

After one forecast realization, let the n x 1 error vector $\underset{\sim}{x}$ generate the matrix statistic $\underset{\sim}{x}\underset{\sim}{x}'$. Thus, given a prior density function, equation 2.19, on $\underset{\sim}{S}$ with prior parameters v^* and $\underset{\sim}{\Psi}^*$ and the observed statistic $\underset{\sim}{x}\underset{\sim}{x}'$, then the posterior density function on $\underset{\sim}{S}$ is also Inverted Wishart with posterior parameters.

$$v^{**} = v^* + 1 \quad (2.23)$$

$$\underset{\sim}{\Psi}^{**} = (v^* \underset{\sim}{\Psi}^* + \underset{\sim}{x}\underset{\sim}{x}')/v^{**} \quad (2.24)$$

Thus equation 3 can now become

$$\underset{\sim}{W} = \underset{\sim}{\Psi}^{**}\underset{\sim}{E} \qquad (2.25)$$

Estimation of the Prior Density Function

Perhaps one of the easiest ways of encoding prior belief in terms of the Inverted Wishart density function is to reduce the problem to that of assessing the marginal inverse gamma distributions for the n diagonal variances of $\underset{\sim}{S}$ and the $\frac{1}{2}n(n-1)$ intercorrelation coefficients. Modal values could be assessed and use be made of equation 2.22, or expectations via equation 2.21. For consistency the parameter v^* should be the same for each of the n inverse gamma distributions. Alternatively, v^* can be considered to represent the hypothetical number of forecast realizations which would be necessary to develop equivalent empirical data to the individual's subjective evidence. Winkler (1967) discusses hypothetical sample methods of this type in the context of assessing beta parameters.

On the basis of general psychological evidence (c.f. Tversky, 1974) it is to be expected that all these methods will tend to underestimate the full extent of the individual's uncertainty, although there is very little specific experimental work relevant to this particular problem.

Empirical Synthesis From Zero Prior Knowledge

An operational Bayesian method should also allow sensible inference and decision making when a completely uninformative prior density function is incorporated. This may arise either because the decision maker actually has no prior belief at all on the covariance matrix or because he wishes to pursue a Bayesian analysis purely on the basis of empirial evidence.

It is suggested that an appropriate uninformative prior density function would be one in which $v^* = 0$. From equation 2.24, it is clear that this allows the initial $\underset{\sim}{\Psi}^*$ to be eliminated after the first observation and thereby avoids subsequent bias. We require the uninformative prior to give equal weight to each model and thus $\underset{\sim}{\psi}^*$ can be conveniently formulated as the identity matrix. This prior density function with $v^* = 0$ will be improper but can be justified along the lines adopted by Lindley (1965) and Jeffreys (1961) for using a gamma prior with $v^* = 0$ in the analysis of the univariate Normal process.

A problem in using this uninformative prior is the $\underset{\sim}{S}$ becomes singular after the first observation, the implicit correlation coefficient being either +1 or -1. As the inverted Wishart density function is not defined on a singular matrix and can therefore be considered uninformative, it is plausible to adopt the rule of assigning models equal weight whenever $\underset{\sim}{S}$ is singular.

However, even when the matrix does become positive, definite and symmetric, the next few observations can still be subject to what might be called early instability. During these early observations, whilst the matrix remains nearly singular, there is a tendency for very large positive and negative weights to occur. The weights will always sum to unity, but there is no constraint for them to remain positive. Thus, weights 1000 and -999, for example, can occur in a situation of two models.

The problem with this orthodox approach to the uninformative prior is that it does not incorporate any prior judgment on the correlation coefficient. Following Lindley (1965) and Jeffreys (1961), an uninformative prior point estimate for $|\rho|$ should be $\frac{1}{2}$. Unfortunately, it is futile to attempt to incorporate this $\underset{\sim}{\psi}^*$ when $v^* = 0$ as it can be seen from equation 2.24 that no subsequent account of it will be taken on the first observation, which is precisely where it is needed. The answer could be to suspend its introduction until the first observation and at that point multiply all the off-diagonal elements of $\underset{\sim}{\psi}^{**}$ by $\frac{1}{2}$. This seems reasonable in view of the recognized fact that the only information the first observation contains on ρ is an indication of its sign.

A series of simulation experiments have established the efficiency of such a suspended Inverted Wishart uninforamtive prior. Independent forecast errors from a set of independent pseudo-Normal predictive models with prespecifiable variances were sampled over a time series of 30 and synthesized for both the "orthodox" prior and the "suspended" uninformative prior. The results shown in Figure 2.2 for 1000 simulations of 3 forecasting models (with $\sigma = 1, 2 \& 4$) are typical and demonstrate the stabilizing effect of the suspended prior. The experiment models the real use of these Inverted Wishart algorithms, in the sense that at time T all of the previous errors for the three models are used to evaluate posterior parameters and hence the prediction for time T + 1. Then the information at time T + 1 of the sample of errors again revises the parameters and in this way the performance over the first 30 realizations is simulated. The computational algorithms, written in BASIC, are provided in the Appendix.

2.3 A Subjective Probability Synthesis of Forecasts

Compared with the tasks of assessing the prior parameters $\underset{\sim}{\psi}^*$ and v^* in the Inverted Wishart formulation, a synthesis based upon linear weights with a direct subjective probability interpretation would appear very attractive. Furthermore, the minimum variance synthesis implies a quadratic loss function for errors and this may not be appropriate. The usual Bayesian principle is to encode all the available information into the subjective probability distribution and then consider utility (loss function) aspects separately.

Recall Section 1.6 where a subset (A) of the total evidence available to the decision maker is considered to be related to the relative forecasting

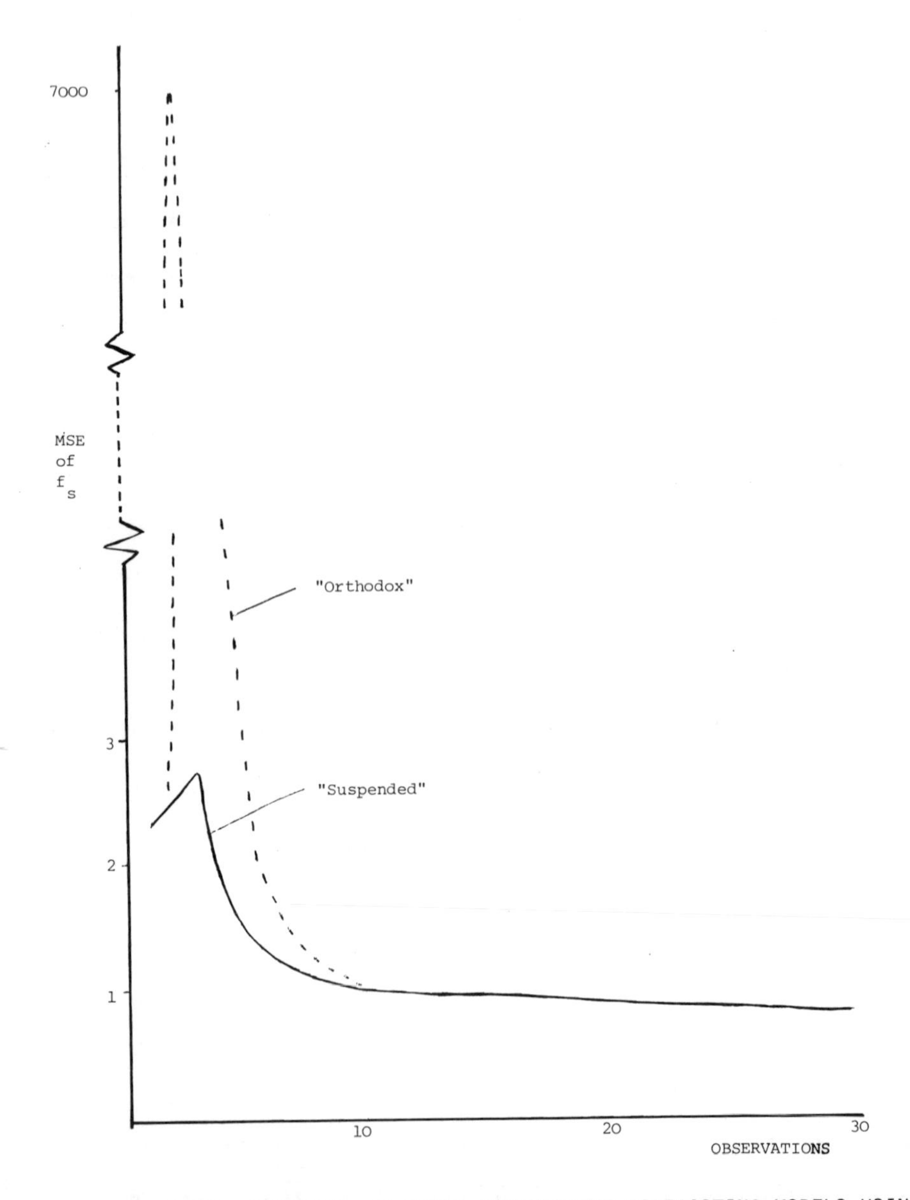

FIGURE 2.2 A LINEAR SYSTHESIS OF THREE INDEPENDENT FORECASTING MODELS USING "ORTHODOX" AND "SUSPENDED" INVERTED WISHART UNINFORMATIVE PRIOR DENSITY FUNCTIONS

ability of each of the n models $f_i(y|m_i)$. A subjective probability distribution can, in principle, be defined over (M) on the basis of (A), i.e. $P(m_i|A)$. Since each forecasting model used to predict y is formulated under its structural assumptions m_i and the relevant subset of (S), the available data, then a synthesized forecast based upon all triples (m_i, s_j, a_k), as advocated in 1.6, can be decomposed as

$$f_s(y|M,S,A) = \sum_{i=1}^{n} f_i(y|m_i,S)P(M_i|A) \qquad (2.26)$$

(the M, S & A are usually suppressed from the formulae by assumption). Point estimates k_i of $P(m_i|A)$ can be taken as linear weights. The vector $\underset{\sim}{k}$ not only satisfies

$$\sum_{i=1}^{n} k_i = 1 \qquad (2.27)$$

as before (equation 2.3), but also

$$0 \leq k_i \leq 1, \qquad (2.28)$$

avoiding the negative weights of $\underset{\sim}{k}^*$ from equation 2.15.

If $\bar{\underset{\sim}{k}}$ is the posterior mean vector of $P(M|A)$, then the posterior mean of the synthesized forecast, given by equation 2.26, is

$$\bar{f}_s = \underset{\sim}{F}'\bar{\underset{\sim}{k}} \qquad (2.29)$$

analogous to equation 2.16 for f_s^*.

If (A) = Ø, i.e. the decision maker has no indication on the relative merits of (M), the appropriate uninformative prior distribution (c.f. Bunn, 1976) is the uniform

$$k_i = 1/n \qquad (2.30)$$

It is worth demonstrating the rationality in decision-analytical terms of synthesizing predictors even in the uninformative situations.

Figure 2.3 displays a decision tree for the problem of selecting one of

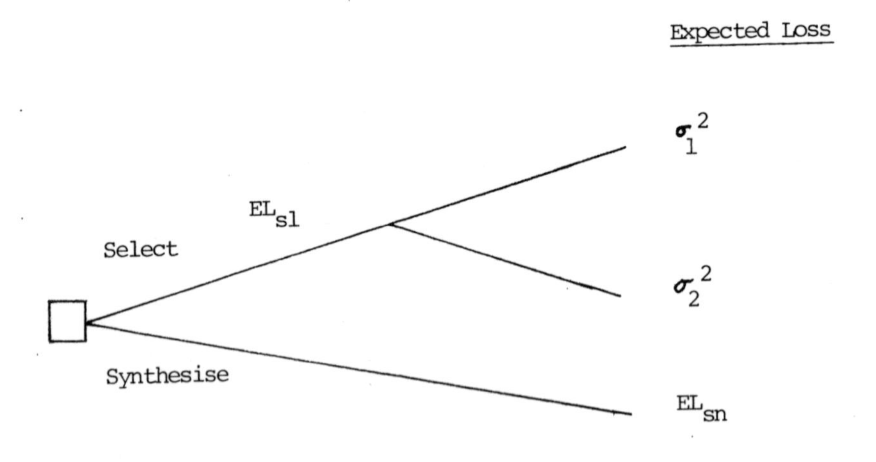

FIGURE 2.3 THE SYNTHESISE OR SELECT DECISION TREE FOR TWO FORECASTING MODELS (σ_1, σ_2) WITH UNIFORM PRIOR PROBABILITIES

two unbiased forecasting models with variances σ_1^2 and σ_2^2, or of synthesizing by means of a linear combination.

Since the decision maker has no information on σ_1 and σ_2, we assume, therefore, that he has no preference between them and would effectively select one at random, if he were to select one at all. Similarly, his synthesis will use the uniform prior probabilities of ½.

Under the usual assumption of a quadratic loss function on forecasting error, it is quite easy to show that the expected loss is proportional to the variance of the model.

The selection of one predictor at random, probability ½, will give an Expected Loss

$$EL_{sl} = \tfrac{1}{2}(\sigma_1^2 + \sigma_2^2) \qquad (2.31)$$

From Equation 2.7, the synthesis with weights of ½ will give an Expected Loss

$$EL_{sn} = \tfrac{1}{4}(\sigma_1^2 + \sigma_2^2 + 2\rho\sigma_1\sigma_2) \qquad (2.32)$$

Thus the Expected Gain from Synthesis

$$EGS = EL_{sl} - El_{sn} \qquad (2.33)$$

$$= \tfrac{1}{4}(\sigma_1^2 + \sigma_2^2 - 2\rho\sigma_1\sigma_2) \qquad (2.34)$$

$$= \tfrac{1}{4}(\sigma_1 - \rho\sigma_2)^2 + (1 - \rho^2)\,\sigma_2^2/4$$

EGS is therefore never negative and only zero is the trivial cases of

Case 1: $\sigma_1 = \sigma_2 = 0$ (2.35)

i.e. each model is a perfect predictor, or

Case 2: $\sigma_1 = \sigma_2$ and $\rho = 1$ (2.36)

i.e. the models are statistically identical.

Thus with no prior information on σ_1, σ_2 and ρ, it is clearly optimal to synthesize the forecasting models. As a typical example, EGS is plotted against ρ, for $\sigma_1 = 1$; $\sigma_2 = 2$, in Figure 2.4.

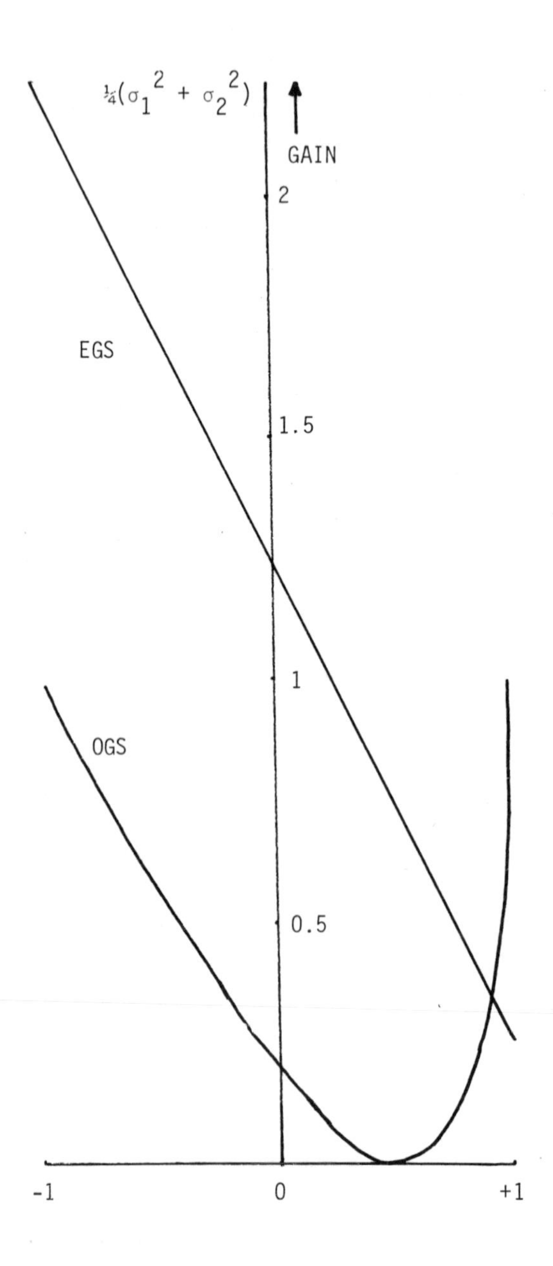

FIGURE 2.4 EGS AND OGS FROM TWO FORECASTING MODELS $\sigma_1 = 1$, $\sigma_2 = 2$ AS A FUNCTION OF ρ

With complete information on σ_1, σ_2 and ρ, at the other extreme, the Optimum Gain from Synthesis, OGS, is given from

$$OGS = EL_{sl}^{*} - EL_{sn}^{*} \tag{2.37}$$

where EL_{sl}^{*} is the expected loss from selecting the minimum variance predictor and EL_{sn}^{*} is the optimum minimum variance synthesis from equation 2.16. Thus, for the example in question, OGS can be derived quite simply from Figure 2.1 and is plotted for comparison also in Figure 2.4.

It is interesting to observe that the EGS with no information and therefore sub-optimal weights is generally greater than the OGS with perfect information. This does not, of course, imply that the extra information is having an adverse effect.

The value of the Expected Loss through synthesis decreases with more precise estimation of the weights, but the net benefit over selection also decreases. This, perhaps, illustrates the power of the decision analysis methodology in dealing with uncertain situations.

The use of an informative probability measure $P(M|A)$, where $(A) \neq \emptyset$, does however pose certain problems of interpretation. Forecasting models are essentially inductive hypotheses and the issues involved with measuring these in subjective probability terms are even more complicated than for straightforward events. This is, therefore, an appropriate point to consider, rather more carefully, the formulation of a meaningful and practical interpretation of subjective probability within the decision-oriented forecasting context.

The Probability of an Inductive Hypothesis

There sometimes appears an almost dogmatic fervor amongst statisticians surrounding the question of whether probability is a degree of belief, a propensity, a relative likelihood or frequency, etc., as if the issue were one of identifying the true predicative quality of probability. However, the mathematical theory of probability can be (and most often is) developed in abstract and it is only when problems of statistical inference arise that a suitable interpretation has to be established. Thus it is suggested that a more open minded approach is to look for the appropriate interpretation of formal probability theory for the problem at hand. In particular, the concern here will be with a suitable interpretation for decision-oriented prediction. Hence, it will be apparent that the statistical reasoning favored by scientists in developing a descriptive theory of Nature will not generally be applicable.

The fundamental abstract theory of probability (sometimes called mathematical or tautalogical) is easily established. Given a σ-field, (E), which can be a set of events or propositions, a probability measure $P(\cdot)$ is defined according to

$$P(E_i) \varepsilon [0,1] \tag{2.38}$$

$$P(E_i) + P(\bar{E}_i) = 1 \tag{2.39}$$

$$P(E_i \cup E_j) = P(E_i) + P(E_j) - P(E_i \cap E_j) \tag{2.40}$$

The decision orientation of the argument clearly implies that consideration must be restricted to the class of interpretations which can admit the incorporation of subjective belief. Thus, the objective, empirical, interpretation of intrinsic probability, propensity and frequency can be excluded from the outset. Their manifest inappropriateness to subjective decision making has been convincingly demonstrated by most of the leading writers of Bayesian inference [c.f. de Finetti (1970); Fishburn (1964); Good, (1965); Jeffreys, (1961); Lindley, (1971); Savage, (1954)].

The standard decision theory approach imputes a subjective probability measure from an individual's revealed preferences amongst a set of gambles.

Savage (1954), Schlaifer (1959) and Anscombe and Aumann (1963) have been influential in the development of this approach. One problem with this interpretation, however, is that it does not facilitate a subjective probability measure on an inductive hypothesis.

Consider in the first place a non-stochastic hypothesis such as "all ravens are black," a proposition dear the the philosophical literature. A rational person will always bet on the falsity of such a proposition because of the impossibility of it's ever being confirmed. A non-black raven may possibly appear, but one could never be certain of having investigated the whole indeterminate population in order to confirm the hypothesis. Thus the imputed probability of any such hypothesis would be zero.

With a stochastic hypothesis such as "data stream (X) follows a linear trend with a random disturbance term," the problem is even more ambivalent as the hypothesis cannot now even be falsified. Any finite number of outliers can be observed without refuting such an inductive hypothesis defined over the full Time-Space domain.

Thus, a pragmatic solution could be to restrict the domain of applicability of the hypothesis and to establish a probability measure on the relative ability of each model to outperform the others belonging to a definite total set of models over a definite number of realizations.

An Outperformance Probability Measure

The motivation for using the outperformance construct is one of decision theory rather than science. The focus of attention on the fixed domain of the appropriate future realizations reflects the precise concern of the decision analyst in the performance of the models. Thus an adaptive, autoregressive model such as one of the Box-Jenkins ARIMA family may be given a relatively high outperformance probability despite the fact that it may contribute much less to a causal explanation of the process than say an econometric model. For short lead times the Box-Jenkins model would be given the higher outperformance probability, but for longer lead times, the econometric model would generally be deemed better. Thus, outperformance probabilities do not reflect the intrinsic scientific relative truth of the models but rather reflect the pragmatic relative ability to perform best over the domain of interest.

The outperformance construct reduces the problem of assigning a probability measure on an inductive hypothesis to one of assessing the expectation (in de Finetti's sense, 1970) for the performance of a set of realizations. Given the next z forecast realizations, the Bayesian expectation for the number of times model i will outperform all others can be estimated in any of the usual ways described in the Appendix applicable to realizable events. A simplex $\underset{\sim}{k}$ of outperformance probabilities can then easily be derived by dividing the set of expectations (E) by z.

$$k_i = E_i/z \qquad (2.41)$$

If $z = 1$, i.e. the next outcome, then the situation reduces to the usual subjective probability assessment task for a realizable event. The interpretation of outperformance has the flexibility of being judgment based but, with the objective generally being to maximize forecast precision, outperformance will usually refer to absolute error. Furthermore, in order that the simplex $\underset{\sim}{k}$ can be described by a Dirichlet density function and revised according to the Bayesian analysis of the multinomial process, it is suggested that the outperformance binary relation should be a strict ordering. This implies that, given the binary relation R, then either xRy or yRx but not both; in other words, model outperformance represents exclusive outcomes and, at any realization, only one model can outperform the set.

If belief in the relative efficacy of each model in the feasible set of models (M) is articulated in terms of the outperformance simplex $\underset{\sim}{k}$ having the Dirichlet density function

$$P(k_1,\ldots,k_n) = [\Gamma(a_1+a_2+\ldots+a_n)/\prod_{i=1}^{n}\Gamma\, a_i](\prod_{i=1}^{n} k_i^{a_i-1}) \tag{2.42}$$

with $\sum_{i=1}^{n} k_i = 1$ and all a_i positive

then the posterior expectation of k after j realizations will be

$$\bar{k}_{i,j} = (a_i + s_{i,j})/(\sum_{i=1}^{n} a_i + j) \tag{2.43}$$

where $s_{i,j}$ denotes the number of times out of j that model i outperformed all other M-1.

If the usual axiomatization of probability in terms of a weak linear order were used, then if two models x and y both equally outperformed the set, by the multinomial theory, $s_{x,1}$ and $s_{y,1}$ would both advance by one. But with the constraint that

$$\sum_{i=1}^{n} s_{1,j} = j \tag{2.44}$$

This would imply that either the posterior expectations do not sum to 1 or that the information base is advanced by more on those occasions in which several models performed equally. Both these situations would be untenable. Hence, the proposed strict ordering or Artificial Exclusivity Devise (AED) implies that whenever a subset X of M performs equally and outperform $\bar{X}$ then each $s_{i,1}$ $(i \in X)$ will advance by the fraction, 1/(number of models in X). It also implies that if the parameters a_i of the prior Dirichlet density function are being assessed by one of the methods of imaginary sample (c.f. Good, 1965) then AED should be used as a conceptual device imposing infinitesimal discrimination between models.

It is important to remain aware that although the time series being forecast may be a non-stationary one, the belief in the relative efficacy of the predictive models should be a stationary probability density function over the defined domain. The only information with implications for the revision of the outperformance probabilities should be the actual realizations themselves; otherwise a start should be made with a fresh prior. Over the domain, it is necessary for the application of the Dirichlet-Multinomial Theory that the ordering of all the bits of information should be irrelevant and exchangeable. The Bayesian procedure is only one of estimation, not adaptation to moving parameters.

Subjective Probability As An Estimate

The limitation of the usual behavioral measure of subjective probability in the organizational context of policy and forecasting have been emphasized in Part 1. Even in the case of individual choice behavior, moreover, this inherently descriptive measure is somewhat inadequate. Insofar as it is imputed from coherent choice behavior, there is no reason to consider it equivalent to a more representative frequentist approach. Yet, there is a temptation to calibrate assessors as good or bad on a frequentist criterion and to evaluate bias in their assessments. Thus, Barclay and Peterson (1973) state:

> If a person gives a 33.1/3% credible interval then, in the long run, 33.1/3% of the true values should fall within the interval, 1/3 above and 1/3 below.

Evidently, in order to be employed in this manner, subjective probability must be considered to be more than just an idiosyncratic decision parameter, but rather an estimate of some underlying true probability. Lottery evaluated subjective probabilities are often criticized for introducing bias [c.f. Winkler, (1976)], yet bias is an empirical construct which should be irrelevant to an interpretation of subjective probability based solely upon measuring a cognitive aspect of an individual.

If, however, subjective probability is treated as an estimate of an underlying logical probability, then principles of rational decision making move away from the self-consistency of Savage's formulation and a normative theory of choice can instead be based upon using such derived subjective probabilities as unbiased estimates of the objective probabilities used in von Neumann and Morgenstern's original (1947) Expected Utility Theory. Such a separation of the probability and utility assessment tasks would be in keeping with some of the implications of policy analysis where it may be more appropriate for different individuals to perform these tasks. Moreover, the need for more explicit formalizations of belief, as for example in the synthesis of forecasts, implies the necessity of some common fundamental logical basis. A set of subjective probabilities cannot be synthesized unless they are believed to be estimates of some common quantity.

However attractive such a separation appears to be, it is still necessary to find an operational interpretation for what is meant by "underlying logical probability." Some conception of the logical implication of the total information base might appear a useful ideal from the Bayesian perspective. This is the sort of conceptual structure Good (1965) uses in developing his hierarchy of probability types. The <u>estimation</u> aspect is only a recognition of our finite information processing ability and inability to formulate a complete set of reasonable beliefs over the whole set of hypotheses and facts in our experience.

To probe deeper into and inquire what purports to the ideal reasonable belief leads into an unhelpful and almost circular area of philosophical inquiry. Popper (1963) and Rescher (1973) advocate a theory of rational belief based upon Tarski's ideas of correspondence and a coherence theory of truth respectively. Quite simply, Tarski's rational belief is a correspondence with facts and Rescher goes on to postulate that this should be expressed in terms of a Many Valued Logic (i.e., a probability measure). The belief structure moreover should satisfy the subjectivist requirement of total coherence, discussed in Section 1.2, but no further help is given to us in constructing this probability measure.

Under a coherent, correspondence theory of belief, however, an ideal logical probability can be postulated which not only embeds the necessary implication of the set of total evidence, but also gives the expectation of conforming to long-run empirical behavior. If the subjective probability assessment task is considered as a human attempt to emulate this ideal, then topics of estimation bias, individual calibration, and algorithms for the improvement of thinking structures advocated in Section 1.2 for policy formulation can be investigated as conceptually relevant.

Two main points have therefore emerged regarding the conceptual framework that a decision-oriented forecaster should adopt in his use of probability theory. It was suggested that the outperformance construct could reduce the problem of assigning a probability measure on a set of predictive models to that of a set of realizable events. Furthermore, it was argued that in assessing the subjective probability of future events, a decision-oriented forecaster should consider his task more as one of estimating an underlying logical implication than of imputing a belief measure from a set of artifical lotteries.

Likelihood Ratio Methods

The posterior model probabilities obtained from the Bayesian analysis of scientific model discrimination (c.f. Section 1.5) have sometimes been used as $P(m_i|A)$ to affect the linear synthesis of equation 2.26. Examples of this Bayesian method are Giesel (1974), Wiginton (1974) and Zellner (1971). The prior probabilities, $P(m_i|A)$, are assessed over the set (M) to represent truth-values of each inductive hypothesis. Although $f_i(y_{t+1}|m_i, S_t)$ represents the predictive distribution of y_{t+1} at time t, after y_{t+1} has been observed at the next instant of time and thus joins an expanded data set S_{t+1}, $f_i\ (y_{t+1}|m_i,\ S_t)$ is also the functional form of the likelihood of model m_i given the observed data, y_{t+1}.

Thus, the posterior probability is given by Bayes Theorem

$$p(m_i|A_{t+1},\ S_{t+1}) = h\ f_i(y_{t+1}|m_i,\ S_t)P\ (m_i|A_t) \tag{2.45}$$

where h is a constant of proportionality to ensure

$$\sum_{i=1}^{n} p(m_i|A_{t+1},\ S_{t+1}) = 1 \tag{2.46}$$

With an initially uninformative prior distribution, the posterior probabilities over (M) clearly reflect the relative likelihood ratio of the models given the data.

Probabilities computed in this way have been used to effect a linear probabilistic synthesis by Wood (1974) on three time-series models of flood frequency and by Baecher and Gros (1975) on three polynomial trending models for geological structures.

It is clear that the interpretation of P(M) is not based upon the relative ability of each model to outperform the rest of the set, but on the possibility of each being the one true model in the set. The notion that (M) contains the one true model is inherent in this formulation. In the analysis of Wood (1974), it becomes quite explicit:

$$f_s(y \mid r, \underset{\sim}{\theta}, \cdot) = \sum_{i=1}^{n} \theta_i f_i(y \mid \cdot) \tag{2.47}$$

$\theta_1, \ldots, \theta_n$ are parameters that take on a value of either or 0 or 1; their values is uncertain. If $\theta_1 = 1$, then model $f_i(y \mid \cdot)$ is the true model. The constraint

$$\sum_{i=1}^{n} \theta_i = 1$$

is imposed, which implies that one and only one model is the true model.

$P(m_i)$ in our notation is then interpreted as $P(\theta_i = 1)$.

The argument developed throughout previous sections has indicated the importance of taking an approach from decision analysis, not scientific discrimination, in formulating the methodology for synthesizing forecasts. Consideration should, therefore, be given to adapting the Bayesian probabilistic scheme of model synthesis to the pragmatic outperformance interpretation of probability discussed previously.

The Outperformance Family of Methods

With two forecasting models, the relative probability that one model will outperform the other over a finite domain can conveniently be assessed as a beta distribution. This relative probability can, of course, be assessed in many of the general ways indicated in Bunn and Thomas (1975), for example as a posterior odds ratio, but the assignment of a beta probability measure has several attractions. Most importantly, it provides the basis for a valid learning process with which to revise the outperformance probability when the models' performance becomes evident. Whether or not one model outperforms the other can be considered to be a Bernoulli variable, and since the beta distribution is a natural conjugate to the Bernoulli process, then the posterior outperformance probabilities will also be beta with a simple modification of the parameters.

It is this dynamic aspect of the beta-Bernoulli process that has been largely reponsible for the recent growth in its applicability. Bierman and Hausman (1970), for example, have utilized it to deal with the credit granting decision and Murray and Silver (1966) in dealing with market share forecasting.

If k is distributed $B(k|a_1, a_2)$ the density function

$$p(k) = [B(a_1, a_2)]^{-1} k^{a_1-1} (1-k)^{a_2-1} \tag{2.47}$$

where $$B(a_1, a_2) = \Gamma a_1 \, \Gamma a_2 \, / \, \Gamma(a_1 + a_2) \tag{2.48}$$

and for k

$$\text{mean} = a_1/(a_1 + a_2) \tag{2.49}$$

$$\text{mode} = (a_1 - 1)/(a_1 + a_2 - 2) \tag{2.50}$$

$$\text{variance} = a_1 \, a_2/(a_1 + a_2)^2(a_1 + a_2 + 1) \tag{2.51}$$

with $k \varepsilon [0,1]$; $0 < a$; $0 < a_2$

With each forecast realization, the new item of data, δ, will represent if F_1 (forecasting model 1) has outperformed F_2, when $\delta = 1$; or not, when $\delta = 0$. Hence δ can be considered to be a Bernoulli variable. After j forecast realizations, and providing the forecaster believes the situation to be stationary, his posterior distribution for k should be

$$B\,(k|a_1 + s_j, a_2 + j - s_j) \tag{2.52}$$

where $$s_j = \sum_{i=1}^{j} \delta_i \tag{2.53}$$

If the situation is deemed by the forecaster not to be stationary, that is, if information other than the forecast realizations has been gained on the relative efficacy of the 2 models, then he should reassess the outperformance measure accordingly.

The posterior mean of distribution 2.9 is given simply from the formula:

$$\bar{k} = (a_1 + s_j) / (a_1 + a_2 + j) \quad (2.54)$$

and the synthesized forecast is accordingly:

$$F = \bar{k} F_1 + (1 - \bar{k}) F_2 \quad (2.55)$$

The argument for using the posterior mean in this context has been discussed in Bunn (1976).

The multivariate analogue of the beta distribution is more commonly known as the Dirichlet distribution, and this is also the natural conjugate for the multinomial process. If the subjective probability (k_i) on the forecasting method i, (F_i), is to be interpreted and assessed as the fraction of z future predictions that it outperforms all other n-1 methods, then the vector ($k_1 \ldots k_n$) can have the Dirichlet density function

$$p(k_1 \ldots k_n) = [\Gamma(a_1 + a_2 \ldots + a_n)/\prod_{i=1}^{n} \Gamma a_i](\prod_{i=1}^{n} k_i^{a_i - 1}) \quad (2.56)$$

with $\sum_{i=1}^{n} k_i = 1$ and all a_i positive.

The posterior mean of k_i, after j realizations, will be given by

$$\bar{k}_{i,j} = (a_i + s_{i,j})/(\sum_{i=1}^{n} a_i + j) \quad (2.57)$$

where $s_{i,j}$ denotes the number of times, out of the j forecast realizations, the method i outperformed all the others.

This follows from the fact that the distribution of each k_i is marginally beta, $B(k_i|a_i, \bar{a}_i)$ where

$$\bar{a}_i = \sum_{j \neq i} a_j \quad (2.58)$$

Although expressed here as n-variate, as the k_i only define an (n-1) dimensional simplex, the distribution has only (n-1) independent variables, and for this reason is often referred to as the (n-1) multivariate beta distribution.

One problem with the straight-forward Dirichlet process is that those relative probabilities in the subset (G) of forecasts which are outperformed at one instance do not undergo any revision despite the extent to which forecasts may be considered to outperform each other within (G). It would seem better, therefore, to assess and revise the relative probability of forecasting method i to forecasting method (i + 1) according to the 2-model beta process. With such a relative subjective probability denoted by p_i, and hence the weights for the pairwise combination of methods i and (i + 1) denoted by $\bar{p}_i$ and $(1 - \bar{p}_i)$, the normalized subjective probability weights, reconciled over the whole set, are easily obtained

$$\text{with} \quad \bar{k}_1 = \bar{p}_1/h \tag{2.59}$$

$$\text{and generally} \quad \bar{k}_i = \bar{p}_1/h \prod_{j=1}^{i-1} [(1-\bar{p}_j)/\bar{p}_j] \tag{2.60}$$

where $h = \sum_{i=1}^{n} \bar{k}_i$ for normalization.

This pairwise beta decomposition is a special case of the generalized Dirichlet distribution described by Connor and Mosimann (1969) and James (1972).

This generalized Dirichlet formulation still does not make full use, however, of a complete outperformance ordering of the forecasting models. Suppose there were 3 models initially given the diffuse prior (1,1; 1,1). After the first realization, furthermore, suppose that the outperformance ranking was 2,1,3. Hence, the posterior generalized Dirichlet would be

(1,2; 2,1) giving the linear weighting factors ¼, ½, ¼. No account has been taken of the fact that model 1 outperformed model 3.

In an attempt to take into account a full linear outperformance ordering of the models, the use of the matrix beta distribution could be considered. The matrix beta is a natural conjugate to matrix multinomial process. Lee, et al. (1968) and Martin (1967) have examined the matrix beta distribution as an aid to estimating transition probabilities in a Markov process. Ezzati (1974) has recently applied such a model to brandswitching behavior.

The conjugate prior to a transition matrix is more strictly speaking, a column vector of Dirichlet distributions, and because the matrix-beta distribution that will be defined here is slightly different, I wish to make this distinction and refer to the transition matrix prior as the matrix Dirichlet.

The conjugate prior density function for the elements of the ith row of the transition probability matrix is the n-variate Dirichlet density function.

$$f_i(p_{i1} \; . \; . \; . \; p_{in}) = [\Gamma \; (\sum_{j=1}^{n} \; a_{ij}) / \prod_{j=1}^{n} (\Gamma \; a_{ij})] \; \Gamma_{j=1}^{n} \; p_{ij}^{a_{ij}-1} \tag{2.61}$$

The marginal distribution of p_{ij} is $B(p_{ij} | a_{ij}, \bar{a}_{ij})$ where

$$a_{ij} = \sum_{j \neq i}^{n} a_{ij} \tag{2.62}$$

If the rows of the matrix $\underset{\sim}{P}$ are independent, there are n such pdf's as equation 2.61 and the joint pdf for all p_{ij} is

$$f(p) = \prod_{i=1}^{n} f_i \; (p_{ij}, \; . \; . \; . \; p_{in}) \tag{2.63}$$

The matrix beta distribution is to be specified simply as a spatial array of beta densities. The probability (k_{ij}) that model j outperforms model i will be assessed as a beta variable $B(k_{ij} | a_{ij}, a_{ji})$, $(i \neq j)$. This is clearly just a more complete evaluation of the pairwise beta approach in the generalized Dirichlet formulation outlined previously. The matrix $\underset{\sim}{K}$ has each element distributed beta subject to the constraint

$$k_{ij} = 1 - k_{ji}, \ (i \neq j) \tag{2.64}$$

in contrast to the matrix Dirchlet $\underset{\sim}{P}$ which has each element distributed beta subject to the constraint

$$\sum_{j=1}^{n} p_{ij} = 1 \tag{2.65}$$

The reason that the matrix Dirichlet formulation cannot be used is that the columns do not relate to exclusive outcomes. The outperformance of i by j, or (j + 1), are not exclusive outcomes, nor can they be constructed to be so by using the artificial exclusivity device employed previously.

The AED was used to ensure

$$(i > j)_{\wedge} (j > i) = \emptyset \tag{2.66}$$

where > is the outperformance binary relation for dichotomous outcomes. In other words, models i and j could not outperform each other; the condition of identical performance if the usual weak partial order were used to axiomatize probability. Instead, if the performance of two or more models were considered identical, then they would be conceptualized as strictly outperforming each other 1/n times (where n is the number of identical models). In the matrix beta case, however, we must allow $[j > i]_{\wedge}[(j + 1) > i]$ to exist by virtue of the postulated complete ranking of outperformance.

It is still worthwhile to preserve AED in the sense of equation 2.25 if only because it allows a sensible interpretation of the diagonal elements of $\underset{\sim}{K}$. k_{ii} will represent the case in which model i is outperformed by no other j, i.e., it outperforms all the models. Thus, the diagonal k_{ii} will represent the Dirichlet variables described previously.

Insofar as the matrix-beta is defined as an array of beta densities with a Dirichlet diagonal, the posterior mean of $\underset{\sim}{K}$ is easily seen to be given by

$$\bar{k}_{ijn} = \frac{a_{ij} + s_{ijn}}{a_{ij} + a_{ji} + n} \tag{2.67}$$

for $i \neq j$ with s_{ijn}, denoting the number of times model j has outperformed model i in n realizations (with AED).

Hence

$$\bar{k}_{jin} = \frac{a_{ji} + s_{jin}}{a_{ij} + a_{ji} + n} \tag{2.68}$$

where $$s_{ijn} + s_{jin} = n \tag{2.69}$$

and a_{ij}, a_{ji} denote the prior beta parameters of $B\ (k_{ij} | a_{ij}, a_{ji})$ and $B\ (k_{ji} | a_{ji}, a_{ij})$

$$\bar{k}_{ii} = \frac{a_{ii} + s_{iin}}{\bar{a}_{ii} + n} \tag{2.70}$$

where s_{iin}, denotes the number of times model i outperformed all others in the n realizations. Hence

$$\sum_{i=1}^{n} s_{iin} = n \tag{2.71}$$

and $$\bar{a}_{ii} = \sum_{j \neq i}^{n} a_{jj} \tag{2.72}$$

with a_{ii}, a_{jj} the prior Dirichlet parameters.

Given the posterior mean matrix $\bar{\underset{\sim}{K}}$ the problem is still to extract a vector of outperformance probability weights. It is suggested that the rows of $\bar{\underset{\sim}{K}}$ are first normalized. The subsequent matrix will then have the properties of a transition matrix and will be referred to as the outperformance probability matrix $\underset{\sim}{Q}$.

$$q_{ij} = \bar{k}_{ij} / \sum_{j=1}^{n} \bar{k}_{ij} \tag{2.73}$$

The steady state eigenvector of $\underset{\sim}{Q}$ can then be evaluated and, as the expected steady state posterior mean vector of the outperformance matrix beta distribution, can be used as the linear weighting vector to synthesize the n models.

Evaluation of the Outperformance Family of Methods

The methods developed above are essentially sophistications of a common theme. A set of forecasting models are synthesized linearly by means of a probability vector which is meaningfully valid and based upon a pragmatic outperformance interpretation. Furthermore, this probability vector can be readily updated according to a rational process of inductive inference and learning. This is because outperformance information is formalized as a n-variate Bernoulli process and the outperformance probability distribution assessed as a conjugate density function.

Outperformance has the additional flexibility of being judgment based. Thus, whether or not one model outperforms another can be assessed by the decision maker; usually, he will be concerned only with absolute error and interpret outperformance soley on this criterion. There are certain circumstances however, where even predictive accuracy has to be assessed subjectively. In situations where forecasts tend to be self-fulfilling, for example the case of monopolistic corporations forecasting behavior in their own markets and making corresponding changes in strategy, or governments adapting policy to their economic projections, it is quite possible that the forecast with the largest "error" may well have been the best predictor of the future had the contingent regulatory action not been taken.

Insofar as outperformance data is not measured according to any metric, but as a zero-one variable in the Dirichlet, a partial ranking in the generalized Dirichlet and a full ordinal ranking in the matrix beta formulations, then there is some loss of optimality in this family of methods. The optimal forecast will clearly consist of a synthesis of models based upon minimizing a loss function defined upon some psychometric outperformance score. With the measurement problems involved in such a formulation projecting its implementability into the realms of the future, and because it seems, therefore, pragmatically optimal at this stage to investigate the characteristics of a first order synthetic approach which appears both theoretically and operationally attractive, consideration will be limited to methodologies dealing with ordinal outperformance data.

As a strong, and perhaps unfair, yardstick with which to evaluate the linear outperformance family, the performance will, however, be compared to that of an optimal synthesis based upon the simple metric outperformance measure of forecast error. This is suggested as somewhat unfair because the fundamental rationale of the outperformance measure is that it is a subjective interpretation of performance and can therefore take into consideration other factors than just absolute error. But since the size of the error will in any case generally be a significant determinant of performance, it is important to see how the linear outperformance family measures up to the optimal synthesis based solely on this factor.

Hence, given two unbiased point estimates (posterior means) from models 1 and 2, F_1 and F_2 and the actual realization A, model 1 will be considered to "outperform" model 2 if

$$\left|F_1 - A\right| < \left|F_2 - A\right| \tag{2.74}$$

The way in which the Dirichlet, generalized Dirichlet and matrix beta make progressively more use of a linear outperformance ranking can be seen

	1	2	3	4	5	6
1	D	G				
2	G	D	G			
3		G	D	G		
4			G	D	G	
5				G	D	G
6					G	D

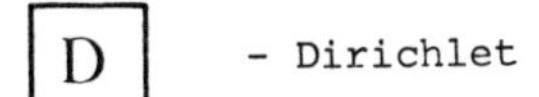

- Dirichlet

G - Generalised Dirichlet

Each cell represents k_{ij}

FIGURE 2.5 OUTPERFORMANCE MATRIX UTILISATION FOR SIX FORECASTING MODELS

from their utilization of the full outperformance matrix $\bar{K}$, Figure 2.5. The matrix beta approach makes full use of each element, the Dirichlet approach only makes use of the diagonal while the generalized beta approach uses just the two sub-diagonals immediately parallel to the diagonal vector.

Other Bayesian variations are possible on this theme, but are not considered to be as practicable as this family. Good (1967), for example postulates a linear combination of Dirichlet distributions in the context of multinomial hypothesis testing. This involves the use of a higher type of probability density function on the Dirichlet parameters. Whether or not this is necessary for estimating support for a null hypothesis is itself controversial (c.f. Lindley's contribution to the discussion of the paper by Good, 1967) and certainly in the context of the estimation of linear outperformance probabilities, it would seem to incur needless conceptual and assessment difficulties.

Simulation Experiments

An exactly analogous set of simulation experiments to those used previously to examine the stability of the minimum variance optimal Bayesian syntheses has been used to compare these three members of the outperformance family of methods. The results shown in Figure 2.6 for σ_i = 1.1, 1.2, 1.2, 1.4, 1.5, 1.6 are typical and indicate the superiority of the matrix beta method. The Dirichlet and Generalized Dirichlet method are prone to a certain amount of "early instability" insofar as they tend to overact to the diagnosticity of the first few early observations.

Figure 2.7 with σ_i = 1.1, 2.1, 1.3 is typical of the results comparing the matrix beta method with the optimal minimum variance synthesis with the suspended inverted Wishart prior. The interesting effect is again the apparent superior stability of the matrix beta. This was found to depend

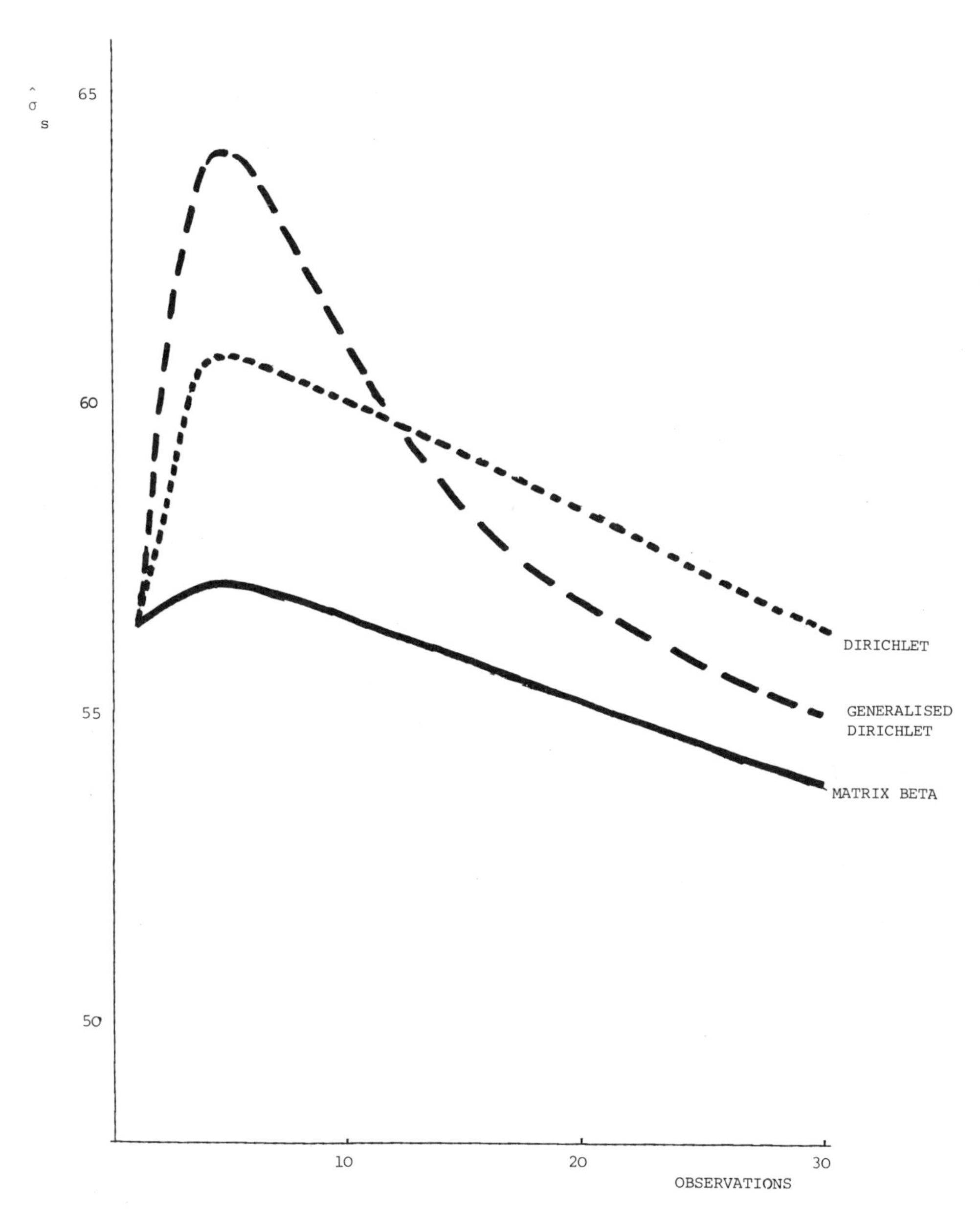

FIGURE 2.6 A COMPARISON OF THE DIRICHLET, GENERALISED DIRICHLET AND MATRIX BETA SYNTHETIC PROCEDURES

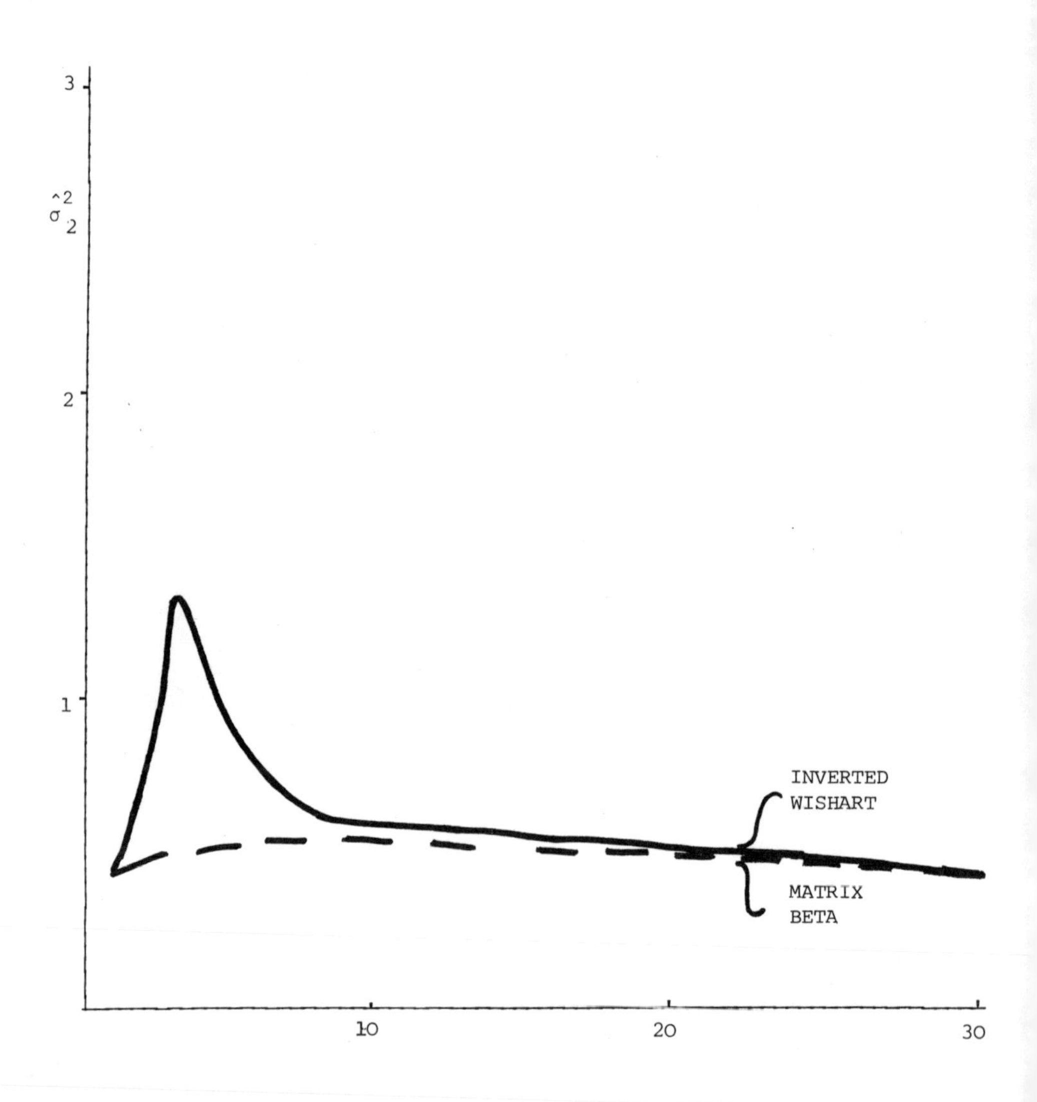

FIGURE 2.7 A COMPARISON OF THE INVERTED WISHART AND MATRIX BETA SYNTHETIC PROCEDURES

upon the number of models and the level of discrimination between them. For example, with $\sigma_i = 1,2,4$ about 10 observations were found necessary before the optimum method became superior compared with the 30 in Figure 2.7. Computational algorithms, written in BASIC are provided in the Appendix.

It should be recognized, however, that an outperformance measure under the absolute error criterion will not necessarily converge to the optimal minimum variance weights. In Figure 2.7, for example, the Inverted Wishart will converge to 0.476 compared with 0.477 for the matrix beta method. This bias will again depend upon the number of models and the level of discrimination between them. For example, with $\sigma_i = 1,2,3$ the Inverted Wishart will converge to a synthesized σ_s of 0.87 compared with 0.94 for the matrix beta.

The above results would therefore tend to suggest that if prior opinion is quite informative and the optimal minimum variance synthesis is required then the Inverted Wishart procedure of Section 2.2 should be used. If prior opinion is, however, relatively uninformative, then the matrix beta will give better results in the short term even under an absolute error interpretation of outperformance.

2.4 Final Comment on Methodology

The methods based upon an optimum minimum variance synthesis (2.2) and a subjective probability simplex on the models (2.3) are not strictly comparable and should not in principle be evaluated as alternatives. The minimum variance procedure is optimal only under the assumption of a quadratic loss function. Furthermore it does not explicitly admit model uncertainty. Each forecasting model is assumed equally valid, although the precision of the parameters is reflected in the optimal linear weights. The subjective probability synthesis, however, is aimed principally at taking explicit account of model uncertainty in the thorough Bayesian Decision Analysis spirit. There is no reason to expect a method based upon a coherent formulation of the full information set to coincide exactly with the optimum under the special criterion of minimum variance.

Insofar as variance will generally decrease as the information content of a probability density function increases, we would not, however, expect the two methods to differ very significantly, and this in fact turned out to be the case as demonstrated in the previous section.

Furthermore, it is a very rare time-series for which the forecasting model is stationary over enough data for the estimated linear weights to become close to their optima. Thus greater importance in practice should, perhaps, be attached to the properties of the methods over the early items of data, and, in particular, to ways of dampening the tendency of many methods to over-react to the diagnosticity of the first few observations.

PART III: APPLICATIONS

Practical case studies cannot provide a justification for a particular forecasting model but they can indicate its feasibility. The case studies discussed here are:

3.1 Forecasting U.K. Output of Gas, Electricity and Water

3.2 Forecasting Hawaiian Tourists

3.3 Forecasting the Sales of a Company in the Home Furnishings Industry

3.4 Forecasting Canadian Consumer Expenditure

3.5 Forecasting the U.K. Economy

3.1 Forecasting U.K. Output of Gas, Electricity and Water

Bates and Granger (1969) and Dickinson (1973) illustrated their discussion of the combination of forecasts by means of a case study which used two simple time series predictors, a linear and an exponential, on the 1948-1965 U.K. output indices for Gas, Electricity and Water, as provided by the Central Statistical Office. As the purpose of case study examples should be to indicate the feasibility of a forecasting method and not a justification of its optimality, the same series was used in Bunn (1975) to illustrate the Bayesian Outperformance methodology. The series has become, it seems, a standard precedent although it is not a particularly ideal example of the power of synthesis as the linear predictor rapidly becomes significantly inferior.

Table 3.1 displays the results of using the beta outperformance method of Section 2.3 with a diffuse prior, $B(k|1,1)$ and an informative one, $B(k|5,50)$. The informative prior is reasonable *a priori* on the general evidence of economic time series, particularly over this period in the U.K., following exponential growth patterns (witness the frequent use of the

Year	Actual Index	Linear Forecast	Exponential Forecast	B(k| 1,1) $\bar{k}$	Forecast	B(k| 5,50) $\bar{k}$	Forecast
1948	58						
1949	62						
1950	67	66.0	66.3	0.50	66.15	0.09	66.27
1951	72	71.3	71.9	0.33	71.70	0.09	71.85
1952	74	76.5	77.4	0.25	77.18	0.09	77.32
1953	77	79.2	80.3	0.40	79.86	0.11	80.19
1954	84	81.9	83.2	0.50	82.55	0.12	83.05
1955	88	89.0	88.6	0.43	88.77	0.12	88.65
1956	92	91.6	93.7	0.38	92.90	0.11	93.46
1957	96	96.0	98.5	0.45	97.37	0.13	98.18
1958	100	100.2	103.2	0.50	101.70	0.14	102.77
1959	103	104.3	107.8	0.55	105.87	0.16	107.25
1960	110	108.1	112.1	0.59	109.74	0.17	111.42
1961	116	112.9	117.4	0.62	114.61	0.18	116.58
1962	125	118.0	123.3	0.57	120.28	0.18	122.35
1963	133	124.2	130.2	0.53	127.02	0.18	129.14
1964	137	130.9	137.8	0.50	134.35	0.17	136.60
1965	145	137.0	145.0	0.47	141.24	0.17	143.63
Mean Squared Error		16.4	5.3	7.3		5.1	

TABLE 3.1 OUTPUT INDICES FOR GAS, ELECTRICITY, AND WATER

logarithmic transformation in the econometric investigations). Thus, if the forecaster believed that the exponential model will outperform the linear model 50 out of 55 times, he would get a smaller MSE by synthesizing the two according to this prior belief rather than selecting the exponential one exclusively. Moreover, if his state of ignorance were such that he had no information to discriminate between them, it would still be preferable to synthesize them according to a diffuse prior than take a 50-50 gamble on selecting a linear predictor with a MSE of 16.4 or an exponential predictor with a MSE of 5.3. Such a random selection would produce a MSE of 10.85 compared with the uninformative prior beta synthesis of 7.3.

Hence this application vindicates the methodology of using prior belief on relative forecast efficacy, not to select out the best model and disregard the others, but to synthesize the models according to this belief as articulated in an outperformance probability measure.

3.2 Forecasting Hawaiian Tourists

Guerts and Ibrahim (1975) fitted a Box-Jenkins predictor and a double exponential smoothing model to the number of tourists visiting Hawaii each month. They then ran these two forecasting models on a further 24 months data and concluded that, since each model had a similar forecast error variance, the exponential smoothing model should be recommended purely on cost-effective grounds.

Table 3.2 shows the output from using the program SYNF/6 (see Appendix 1) which effects a linear synthesis using the optimum minimum variance procedure of 2.2, with the 'suspended' uninformative Inverted Wishart prior. The forecast error variance is evidently reduced from about 2.5M for each model to approximately 0.55M for the synthesized predictor. This quite dramatic reduction in the error variance is attributable to the

TABLE 3.2: OUTPUT OF SYNF/6 APPLIED TO THE HAWAII TOURIST PREDICTORS OF GUERTS & IBRAHIM

```
INPUT THE NUMBER OF MODELS
?2
INPUT THE MATRIX OF PRIOR INVERTED WISHART PARAMETERS
?1,1,1,1
INPUT THE DEGREES OF FREEDOM
?0
INPUT THE LENGTH OF THE TIME SERIES
?24

MODEL          FORECAST          WEIGHT
 1             -1883              .5
 2              375               .5

ACTUAL          0
COMBINED       -754


MODEL          FORECAST          WEIGHT
 1              14                .112395
 2             -336               .887605

ACTUAL          0
COMBINED       -296.662


MODEL          FORECAST          WEIGHT
 1              1251              .135392
 2             -2093              .864608

ACTUAL          0
COMBINED       -1640.25


MODEL          FORECAST          WEIGHT
 1             -819               .484815
 2             -1539              .515185

ACTUAL          0
COMBINED       -1189.93


MODEL          FORECAST          WEIGHT
 1              447               .53765
 2              140               .46235

ACTUAL          0
COMBINED        305.059
```

TABLE 3.2 (Continued)

MODEL	FORECAST	WEIGHT
1	72	.531908
2	644	.468092
ACTUAL	0	
COMBINED	339.749	

MODEL	FORECAST	WEIGHT
1	799	.543589
2	628	.456411
ACTUAL	0	
COMBINED	720.954	

MODEL	FORECAST	WEIGHT
1	162	.536192
2	-822	.463808
ACTUAL	0	
COMBINED	-294.387	

MODEL	FORECAST	WEIGHT
1	1902	.552618
2	-946	.447382
ACTUAL	0	
COMBINED	627.857	

MODEL	FORECAST	WEIGHT
1	-30	.483165
2	1247	.516835
ACTUAL	0	
COMBINED	629.999	

MODEL	FORECAST	WEIGHT
1	1074	.512552
2	275	.487448
ACTUAL	0	
COMBINED	684.529	

TABLE 3.2 (Continued)

MODEL	FORECAST	WEIGHT
1	-427	.493028
2	-1454	.506972
ACTUAL	0	
COMBINED	-947.66	

MODEL	FORECAST	WEIGHT
1	-2629	.526508
2	-692	.473492
ACTUAL	0	
COMBINED	-1711.85	

MODEL	FORECAST	WEIGHT
1	-1697	.425482
2	1263	.574518
ACTUAL	0	
COMBINED	1368.03	

MODEL	FORECAST	WEIGHT
1	-2545	.425736
2	4269	.574264
ACTUAL	0	
COMBINED	1368.03	

MODEL	FORECAST	WEIGHT
1	-39	.531649
2	49	.46851
ACTUAL	0	
COMBINED	2.21489	

MODEL	FORECAST	WEIGHT
1	914	.531651
2	-969	.468349
ACTUAL	0	
COMBINED	32.0993	

TABLE 3.2 (Continued)

MODEL	FORECAST	WEIGHT
1	1331	.530991
2	-1241	.469009
ACTUAL	0	
COMBINED	124.709	

MODEL	FORECAST	WEIGHT
1	-665	.527724
2	416	.472276
ACTUAL	0	
COMBINED	-154.47	

MODEL	FORECAST	WEIGHT
1	2478	.526044
2	-2179	.473956
ACTUAL	0	
COMBINED	270.785	

MODEL	FORECAST	WEIGHT
1	3307	.515625
2	-2381	.484375
ACTUAL	0	
COMBINED	551.875	

MODEL	FORECAST	WEIGHT
1	2764	.495161
2	-3861	.504839
ACTUAL	0	
COMBINED	-580.561	

MODEL	FORECAST	WEIGHT
1	57	.514657
2	-160	.485343
ACTUAL	0	
COMBINED	-48.3195	

TABLE 3.2 (Continued)

MODEL	FORECAST	WEIGHT
1	-781	.51471
2	602	.48529

ACTUAL 0
COMBINED -109.844

MSE MODEL 1 2.34393E+06
MSE MODEL 2 2.57551E+06
MSE COMBINED 554470.

POSTERIOR INVERTED WISHART PARAMETER MATRIX

2.34393E+06	-1.69116E+06
-1.69116E+06	2.57551E+06

DEGREES OF FREEDOM 24

DONE

relatively high degree of negative correlation between the predictors, as examination of the posterior parameter matrix at the foot of Table 3.2 demonstrates.

A synthesis using the outperformance methodology of 2.3 gives a MSE of about 0.4M, again using an uninformative prior $B(k|1,1)$. The posterior beta parameters turned out to be (13,13).

It is only fair to point out, however, that it may well be the case that the optimal Box-Jenkins predictor has not been completely identified in the first place by Guerts and Ibrahim. The exponential smoothing model is also a member of the Box-Jenkins ARIMA class of predictors and, furthermore, it can easily be shown that a linear combination of ARIMA predictors is also a member of the ARIMA family. Thus, even our synthesized predictor is, strictly speaking, a Box-Jenkins predictor and clearly a more optimal one than that originally identified.

3.3 Forecasting the Sales of a Company in the Home Furnishings Industry

In 1971, Parker and Segara, published an article in the Harvard Business Review entitled, "How to get a Better Forecast," which attempted to show how the sales of a particular company in the home furnishings industry could be predicted more accurately by a casual model than a time-series extrapolation. Based upon the data shown in Table 3.3, a regression model was estimated which predicted sales according to the equation:

$$S_t = a + b\,S_{t-1} + c\,H_{t-1} + d\,I + e\,T + \varepsilon \qquad (3.1)$$

The performance of this model was then compared with that obtained by extrapolating a five-year moving average. Figure 3.1 illustrates the performance of these two models. With the MSE for the regression model being

Year	Housing Starts (II) (thousands)	Disposable Personal Income (I) ($ billions)	New Marriages (M) (thousands)	Company Sales (S) ($ millions)	Time (T)
1947	744	158.9	2,291	92.920	1
1948	942	169.5	1,991	122.440	2
1949	1,033	188.3	1,811	125.570	3
1950	1,138	187.2	1,580	110.460	4
1951	1,549	205.8	1,667	139.400	5
1952	1,211	224.9	1,595	154.020	6
1953	1,251	235.0	1,539	157.590	7
1954	1,225	247.9	1,546	152.230	8
1955	1,354	254.4	1,490	139.130	9
1956	1,475	274.4	1,531	156.330	10
1957	1,240	292.9	1,585	140.470	11
1958	1,157	308.5	1,518	128.240	12
1959	1,341	318.8	1,451	117.450	13
1960	1,531	337.7	1,494	132.640	14
1961	1,274	350.0	1,527	126.160	15
1962	1,327	364.4	1,547	116.990	16
1963	1,469	385.3	1,580	123.900	17
1964	1,615	404.6	1,654	141.320	18
1965	1,538	436.6	1,719	156.750	19
1966	1,488	469.1	1,789	171.930	20
1967	1,173	505.3	1,844	184.790	21
1968	1,299	546.3	1,913	202.700	22
1969	1,524	590.0	2,059	237.340	23
1970	1,479	629.6	2,132	254.930	24

Source: Statistical Abstract of the United States (Washington, Bureau of the Census).

Note: Company sales and disposable per-capita income have been adjusted for the effect of inflation and appear in constant 1959 dollars.

TABLE 3.3 DATA FOR THE REGRESSION MODEL OF PARKER & SEGURA (1971)

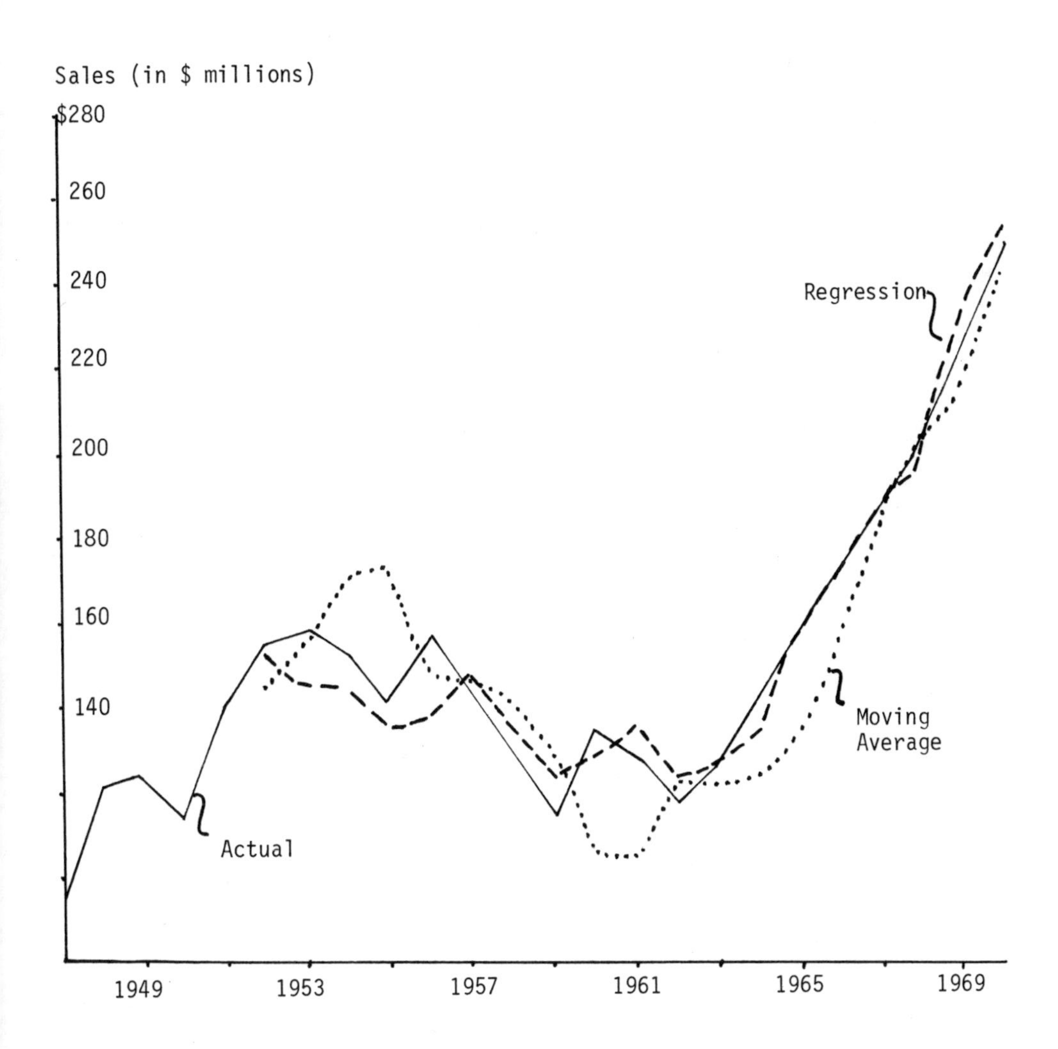

FIGURE 3.1 RELATIVE PERFORMANCE OF THE REGRESSION & MOVING AVERAGE FORECASTING MODELS FOR PARKER & SEGURA (1971)

62 compared with 217 for the time-series extrapolation, it was concluded that such sales forecasts could be better predicted by the causal model.

However, there appeared to be a sufficient amount of independence in the two hypotheses to suggest integrating them together via outperformance probability weights. With only two models, the Dirichlet, Generalized Dirichlet, Matrix Beta, or the plain Beta procedures will give the same results. Two such syntheses were attempted, first with a diffuse prior and then with prior beta weights (25,5). The diffuse prior gave a MSE of 80, the informative prior, a MSE of 60. With an R^2 of 0.95 for the regression model, and in view of the fact that it already incorporates a trend factor, such an informative prior weighting could be considered reasonable.

Thus, if the objective is to obtain the best forecast in terms of accuracy, it is clearly worthwhile to combine these two models according to a subjective probability weighting which is adaptively revised according to a beta-Bernoulli process. Furthermore, on decision analytic grounds, it is still more sensible even with no preference for the models to form a synthesis.

3.4 Forecasting Canadian Consumer Expenditure

In 1970, the Economic Council of Canada commissioned a study to develop a forecasting method applicable over a large set of disaggregated consumer expenditures. The approach should be the same for each, with variation only in the parameters of the model. Schweitzer (1970) recommended that the Houthakker Taylor model would be appropriate.

Support for the H.T. model was by the common econometric device of comparing its performance with two other naive models. The results are shown in Table 3.4. The Naive I model assumed that the per capita real consumption of each item will grow in the 3 year period by the same per-

Item	Actual	H-T*	Naive I	Naive II
Food Purchased at Retail	+ 6.8	+10.8	+ 2.5	+10.3
Food Produced & Consumed on Farms	- 0.2	-24.6	-12.8	+10.3
Other Food	+ 9.3	- 9.8	+ 9.8	+10.3
Nonalcoholic Beverages	+10.3	+ 7.6	+ 9.0	+10.3
Alcoholic Beverages	+ 6.4	+11.4	+11.8	+10.3
Tobacco	- 9.8	- 0.9	+ 1.8	+10.3
Men's & Boy's Clothing	+ 2.6	+ 3.8	+ 7.4	+10.3
Women's & Children's Clothing	+ 5.8	+ 9.9	+ 5.1	+10.3
Notions & Piece Goods	+15.9	+10.9	- 5.0	+10.3
Armed Forces Clothing	-15.0	- 7.5	-21.6	+10.3
Dressmaking & Tailoring	+29.3	+ 1.5	- 5.1	+10.3
Footwear	- 2.3	+ 6.3	- 2.2	+10.3
Shoe Repair	+25.0	+ 2.4	- 0.8	+10.3
Luggage & Leather Goods	+12.2	+ 5.4	+ 2.8	+10.3
Jewellery	+ 9.6	+14.6	+15.5	+10.3
Jewellery Repair	-10.9	-27.3	- 1.8	+10.3
Water Charges	- 8.0	+28.5	+14.3	+10.3
Electricity	+27.3	+19.5	+14.0	+10.3
Gas & Other Fuels	+ 4.3	+ 8.0	+ 5.8	+10.3
Gross Rents, Imputed	+12.2	+14.0	+12.2	+10.3
Gross Rents, Paid	+18.5	+12.9	+16.8	+10.3
Imputed Lodging N.E.S.	+13.3	+ 1.8	+27.2	+10.3
Board & Lodging in Universities	+44.3	+51.6	+56.4	+10.3
House Maintenance Repairs	-18.6	-19.8	+ 1.7	+10.3
Furniture, Upholstery & Furniture Repair	- 4.7	+11.8	+ 5.7	+10.3
Carpets & Other Floor Coverings	+15.0	+27.4	+42.8	+10.3
Household Textiles & Other Furnishings	+12.7	+ 9.9	+22.8	+10.3
Household Appliances	- 0.3	+14.8	+24.4	+10.3
Glassware, Tableware & Household Utensils	+12.5	+14.5	+25.5	+10.3
Household Operation Goods	+ 1.3	+16.0	+ 6.4	+10.3
Household Operation Services	- 0.8	+10.8	+ 6.4	+10.3
Medical & Pharmaceutical Products	+36.8	+13.5	+19.0	+10.3
Medical Care & Health Services	+ 8.9	+13.3	+11.1	+10.3
Personal Transportation Equipment, Auto Repairs & Maintenance, Auto Parts & Accessories	+ 5.5	+ 6.8	+23.8	+10.3
Gasoline, Oil & Grease	+ 4.3	+12.3	+ 1.1	+10.3
Operations of Personal Transportation Equipment	+21.1	+23.2	+47.4	+10.3
Purchased Transportation	+ 1.7	+ 5.9	+ 9.1	+10.3
Communications	+20.6	+13.7	+15.3	+10.3

*Houthakker and Taylor

TABLE 3.4 ACTUAL AND FORECASTED PERCENTAGE CHANGES, 1966-1969 IN ACTUAL PERSONAL COMSUMER EXPENDITURE

centage as in the previous 3 year period. The Naive II model assumes growth at the same rate as per capita real total consumer expenditure.

Although it seems that each of the two naive models is largely embedded in the H.T. model, as an example of the way in which outperformance probabilities can be established over a domain which is not a time series, and in order to see to what extent the performance of the H.T. model could be improved by integrating with two naive models, the three models were synthesized by means of the Dirichlet procedure.

With a diffuse prior, a considerable reduction of the MSE was obtained from 165 for the H.T. model to 123 for the combined forecast. Also, and surprisingly, the posterior Dirichlet parameters 11, 15, 15 indicated that the naive models outperformed the H.T. method. Thus, using what would seem reasonable informative priors of 7, 2, 1 & 70, 15, 15, MSE's of 127 and 120 were obtained for the combined forecast. With prior Dirichlet parameters of 11, 15, 15 the combined MSE was reduced to 121, although such higher prior weighting to the naive models would be difficult to justify a priori.

The Generalized Dirichlet procedure was also used. These findings support the simulation results suggesting that the Generalized Dirichlet is superior to the Dirichlet in estimating the outperformance probabilities. With a diffuse prior, the combined MSE was 122. What is also interesting is that the posterior beta parameters are (21:19; 21:19), indicating that the H.T. model is after all slightly better on outperformance than Naive I, which is itself slightly better than Naive II: hence the advantage of taking pairwise outperformance rather than absolute outperformance as in the Dirichlet. Using 21:19; 21:19 as prior, the MSE was further reduced to just below 121.

3.5 Forecasting the U.K. Economy

Policy-makers rely more and more upon a wider variety of economic forecasts. Intuitively they assign weightings to each model to obtain their own expectation of the economic future. It was considered important, therefore, to examine the feasibility of formalizing this intuitive process by means of linear outperformance probabilities.

Ball and Burns (1970) attempted an evaluation of the London Business School econometric forecasting model by comparing its performance on 4 sectors over the years 1968 and 1969 with 3 other national models. The results are shown in Table 3.5.

Using the Generalized Dirichlet method and a diffuse prior, a MSE of 3.1, compared with 5.8, 7.3, 6.4 and 2.2 for each of the other models respectively, was obtained. Thus, if prior opinion were equally divided between the four models, it is clearly better to combine with equal weights than to select out one at random.

The Matrix Beta method was also run on this data, but gave a slightly worse combined MSE of 4. Table 3.6 shows the output for this experiment. A synthesis using the Dirichlet method, again with a diffuse prior, gave a synthesized MSE of 3.9.

An absolute error interpretation of outperformance would generally not be recommended in the practical evaluation of econometric forecasts. Advantage can be taken of the fact that outperformance is a subjective ranking. Where a government manages the economy according to its econometric models, action (it is to be hoped) is often taken to steer the economy away from undesirable predictions. Thus it is not strictly valid in these circumstances to compare "forecast" with "actual" in order to rank the models. The model with the largest absolute "error" might have been the

(a) 1968

	1 NIESR February 68	2 TREASURY March 68	3 LBS March 68	4 NIESR May 68	5 ACTUAL April 70
Gross Domestic Product[1]	2.9	3.0	3.5	3.4	3.2
Consumption	-0.4	0.2	0.3	1.2	2.1
Exports	8.8	7.9	8.6	9.9	12.1
Imports[2]	2.1	1.9	1.9	6.9	7.1

(b) 1969

	1 NIESR February 69	2 TREASURY April 69	3 LBS April 69	4 NIESR May 69	5 ACTUAL April 70
Gross Domestic Product[1]	3.3	2.3	3.2	2.2	2.0
Consumption	0.4	-0.2	0.6	0.2	0.4
Exports	6.8	5.5	5.7	4.9	8.3
Imports[2]	1.9	0.8	2.0	2.1	2.2

1. Actual value for GDP is the 'compromise' estimate

2. Treasury estimate includes US military expenditure

TABLE 3.5 COMPARISON OF FORECASTS MADE DURING PERIOD FEBRUARY TO MAY IN 1968 AND 1969

TABLE 3.6: THE MATRIX BETA PROGRAM SYNF/4 ON THE UK ECONOMETRIC FORECASTING DATA

```
INPUT THE NUMBER OF MODELS
?4
INPUT THE OUTPERFORMANCE MATRIX
? 1,1,1,1
??1,1,1,1
??1,1,1,1
??1,1,1,1
INPUT THE LENGTH OF THE TIME SERIES
?8

FORECAST   1      2.9          SSE     8.99998E-02
FORECAST   2      3            SSE     3.99999E-02
FORECAST   3      3.5          SSE     9.00001E-02
FORECAST   4      3.4          SSE     4.00001E-02

COMBINED          3.2          SSE     2.27374E-13
ACTUAL            3.2

FORECAST   1     -.4           SSE     6.34
FORECAST   2      .2           SSE     3.65
FORECAST   3      .3           SSE     3.33
FORECAST   4      1.2          SSE     .85

COMBINED          .337032      SSE     3.10805
ACTUAL            2.1

FORECAST   1      8.8          SSE     17.23
FORECAST   2      7.9          SSE     21.29
FORECAST   3      8.6          SSE     15.58
FORECAST   4      9.9          SSE     5.69

COMBINED          8.84917      SSE     13.6759
ACTUAL            12.1

FORECAST   1      2.1          SSE     42.23
FORECAST   2      1.9          SSE     48.33
FORECAST   3      1.9          SSE     42.62
FORECAST   4      6.9          SSE     5.73

COMBINED          3.8675       SSE     24.125
ACTUAL            7.1

FORECAST   1      3.3          SSE     43.92
FORECAST   2      2.3          SSE     48.42
FORECAST   3      3.2          SSE     44.06
FORECAST   4      2.2          SSE     5.77

COMBINED          2.58289      SSE     24.4647
ACTUAL            2
```

TABLE 3.6 (CONT.)

```
FORECAST   1    .4          SSE    43.92
FORECAST   2   -.2          SSE    48.78
FORECAST   3    .6          SSE    44.1
FORECAST   4    .2          SSE    5.81

COMBINED        .296401     SSE    24.5022
ACTUAL          .4

FORECAST   1    6.8         SSE    46.17
FORECAST   2    5.5         SSE    56.62
FORECAST   3    5.7         SSE    50.86
FORECAST   4    4.9         SSE    17.37

COMBINED        5.53815     SSE    32.13
ACTUAL          8.3

FORECAST   1    1.9         SSE    46.26
FORECAST   2    .8          SSE    58.58
FORECAST   3    2           SSE    50.9
FORECAST   4    2.1         SSE    17.38

COMBINED        1.763       SSE    32.321
ACTUAL          2.2

***FINAL MEAN SQUARED ERRORS***

MSE    1          5.7825
MSE    2          7.3225
MSE    3          6.3625
MSE    4          2.1725
MSE   COMBINED    4.04013

POSTERIOR MATRIX BETA PARAMETERS
 3    4    4    7

 6    2    7    7

 6    4    1    8

 3    3    2    6


DONE
```

best predictor of what would have happened had the contingent regulatory action not been taken.

Thus, while it makes sense to synthesize econometric forecasts in order to formulate coherent judgment, care must be taken in revising the prior parameters upon the evidence of the "track record."

REVIEW

The Coherence Principle of rational belief and action is fundamental to this work. It is necessary for rational decision analysis that the decision-maker's opinion upon the future should be coherent with all the evidence open to him. This implies the necessity of attempting an explicitly coherent synthesis of the set of available forecasting models. Part 2 evaluated possible methods of achieving this and their feasibility was demonstrated in Part 3. The Appendix which follows provides the computational algorithms for all of the outperformance family of probabilistic methods and the minimum variance procedures of Section 2.2. It also contains a more thorough discussion of the issue of estimating subjective probabilities.

There is clearly still much work to be done both in the further development of methodology and in extending the range of applications. The methodology has obvious generality. It can provide the basis for an individual's formulation of belief at the conceptual level of subjective probability assessment or at the more explicit level of hierarchical inference. The individual forecasting models need not be mathematical equations but can be expert opinions and the synthesis will then relate to a treatment of the consensus problem. However, the synthesis of a set of disparate predictors such as one might face in practical problems of technological or social policy formulation is the area where such methodology is expected to make the greatest impact. Essentially all types of corporate decision analysis require a synthesis of information, predictions and opinions. The explicit formalization of this judgmental process is evidently just a natural extension of the basic purpose of modern Decision Analysis in clarifying the underlying rationale of preferences and opinions and providing a structure for what was previously a purely intuitive process.

One final caveat, which although rather obvious, is worth emphasizing. The fact that a set of predictors is going to be combined should not detract from the importance of developing good individual predictors. The effort involved in the construction of a smaller number of highly representative models will repay the analyst not only in forecast performance but also with insight into the process, especially if the alternative effort would be a loose amalgamation of a larger number of relatively crude predictors. Just as Box and Jenkins (1970) advocate the virtues of parsimonious parameterization in time series models, there is evidently an analogous virtue in parsimony of predictors, subject of course to the total coherence in a synthesis of these models. Furthermore, it will be recalled that the optimal set of models for synthesis can in principle be evaluated according to the standard Bayesian prior analysis of the value of information as in Raiffa and Schlaifer (1961).

APPENDIX 1: COMPUTATIONAL ALGORITHMS

The computational algorithms, written in time-sharing BASIC, used in the simulation experiments of Part 2 and the applications of Part 3 are listed here. Included are the Beta, Dirichlet, Generalized Dirichlet and Matrix Beta outperformance probability methods and the minimum variance procedures with and without the "suspended" uninformative prior facility.

SYNF/1: The Beta Outperformance Method

```
 10  DIM A[40],F[40],S[40]
 20  PRINT "LENGTH OF FORECAST SERIES="
 30  INPUT N
 40  FOR I=1 TO N
 50  HEAD A[I],F[I],S[I]
 60  NEXT I
 70  PRINT
 80  PRINT "PRIOR BETA PARAMETERS="
 90  INPUT A1,A2
100  PRINT
110  PRINT "ACTUAL", "COMBINED", "FORECAST 1", "FORECAST 2", "F.1
     WEIGHT"
120  M1=M2=M3=0
130  FOR I=1 TO N
140  W=A1/(A1+A2)
150  C=W*F[I] + (1-W)*S[I]
160  PRINT A[I],C,F[I],S[I],W
170  E1=ABS(F[I]-A[I])
180  E2=ABS(S[I]- A[I])
190  E3=ABS(C-A[I])
200  IF E1>E2 THEN 240
210  IF E1=E2 THEN 260
220  A1=A1+1
230  GOTO 280
240  A2=A2+1
250  GOTO 280
260  A1=A1+.5
270  A2=A2+.5
280  M1=M1+(E3↑2)/N
290  M2=M2+(E1↑2)/N
300  M3=M3+(E2↑2)/N
310  NEXT I
320  PRINT
330  PRINT "FORECAST ONE:  MEAN SQUARED ERROR=";M2
340  PRINT "FORECAST TWO:  MEAN SQUARED ERROR=";M3
350  PRINT "COMBINED 1&2:  MEAN SQUARED ERROR=";M1
360  PRINT
370  PRINT "POSTERIOR BETA PARAMETERS=",A1,A2
380  PRINT
390  PRINT "ALTERNATIVE PRIOR?  IF YES, TYPE 1; IF NO, TYPE 0"
```

```
400  INPUT P
410  P>0 THEN 70
420  END
```

SYNF/2: The Dirichlet Outperformance Method

```
 10  DIM A[40,7],[6],M[6],K[6]
 20  PRINT "LENGTH OF THE FORECAST SERIES="
 30  INPUT N
 40  PRINT "NUMBER OF FORECASTS="
 50  INPUT M
 60  M=M+1
 70  MAT READ A[N,M]
 80  M=M-1
 90  PRINT "DIRICHLET PARAMETERS="
100  MAT INPUT D[M]
110  PRINT
120  PRINT       "HYPOTHESIS", "PARAMETER", "FORECAST", "ACTUAL",
     "COMBINED"
130  MAT M=ZER[M]
140  M1=0
150  FOR I=1 TO N
160  MAT K=ZER[M]
170  C=S=Z=0
180  FOR J=1 TO M
190  S=S+D[J]
200  NEXT J
210  FOR J=1 TO M
220  K=J+1
230  C=C+(D[J]*A[I,K])/S
240  E[J]=ABS(A[I,K] -A[I,1])
250  M[J]=M[J]+(E[J]↑2)/N
260  NEXT J
270  E1=ABS(C-A[I,1])
280  M1=M1+(E1↑2)/N
290  FOR J=1 TO M
300  PRINT J,D[J],A[I,J+1],A[I,1]
310  NEXT J
320  PRINT " "," "," ",A[I,1],C
330  B=1.E+06
340  FOR J=1 TO M
350  B=(B MIN E[J])
360  IF E[J]>B THEN 380
370  K=J
380  NEXT J
390  FOR J=1 TO M
400  IF E[J] <> E[K] THEN 430
410  K[J]=1
420  Z=Z+1
430  NEXT J
440  FOR J=1 TO M
450  D[J]=D[J]+K[J]/Z
```

```
460  NEXT J
470  NEXT I
480  PRINT
490  PRINT "COMBINED FORECAST:  MEAN SQUARED ERROR=";M1
500  FOR I=1 TO M
510  PRINT "FORECAST ";I;" ;  MEAN SQUARED ERROR="; M[I]
520  NEXT I
530  PRINT
540  PRINT "POSTERIOR DIRICHLET PARAMETERS"
550  MAT PRINT D
560  PRINT "ALTERNATIVE PRIOR?  IF YES, TYPE 1; IF NO, TYPE 0"
570  INPUT P
580  IF P>0 THEN 90
590  END
```

SYNF/3: The Generalized Dirichlet Outperformance Method

```
 10  DIM A[40,10],B[9],C[9],M[9],E[9],K[9],W[9],G[9]
 20  DIM A$[3]
 30  PRINT "LENGTH OF FORECAST SERIES"
 40  INPUT N
 50  PRINT "NUMBER OF FORECAST MODELS"
 60  INPUT M
 70  M=M+1
 80  MAT READ A[N.M]
 90  M=M-2
100  FOR I=1 TO M
110  J=I+1
120  PRINT "PRIOR BETA PARAMETERS FOR MODEL ";I;" AGAINST
     MODEL";J
130  INPUT B[I],C[I]
140  NEXT I
150  M1=0
160  M=M+1
170  MAT M=ZER[M]
180  M=M-1
190  FOR I=1 TO N
200  PRINT
210  PRINT "MODEL", "FORECAST", "RELATIVE BETAS", "GENERALIZED
     WEIGHTS"
220  C=0
230  GOSUB 250
240  GOTO 360
250  FOR J=1 TO M
260  W[J]=B[J]/(B[J]+C[J])
270  NEXT J
280  S=K[1]=W[1]
290  M=M+1
300  FOR J=2 TO M
310  K=J-1
320  K[J]=K[K]*(1-W[K])/W[K]
330  S=S+K[J]
340  NEXT J
```

```
350  RETURN
360  B[M]=0
370  C[M]=0
380  FOR J=1 TO M
390  K=J+1
400  G[J]=K[J]/S
410  C=C+G[J]*A[I,K]
420  E[J]=ABS(A[I,K]-A[I,1])
430  M[J]=M[J]+(E[J]↑2)/N
440  PRINT J,A[I,K],B[J];"  ";C[J],G[J]
450  NEXT J
460  PRINT
470  PRINT "ACTUAL  =",A[I,1]
480  PRINT "COMBINED  =",C
490  E1+ABS(C-A[I,1])
500  M1=M1+(E1↑2)/N
510  M=M-1
520  FOR J=1 TO M
530  IF E[J]>E[J+1] THEN 570
540  IF E[J]=E[J+1] THEN 590
550  B[J]=B[J]+1
560  GOTO 610
570  C[J]=C[J]+1
580  GOTO 610
590  B[J]=B[J]+.5
600  C[J]=C[J]+.5
610  NEXT J
620  NEXT I
630  PRINT
640  PRINT
650  PRINT " ","***************FINAL ANALYSIS***************"
660  PRINT
670  PRINT "MODEL","MSE","POSTERIOR","POSTERIOR WEIGHTS"
680  PRINT " "," ","RELATIVE BETAS"
690  GOSUB 250
700  FOR J=1 TO M
710  G[J]-K[J]/S
720  PRINT J,M[J],B[J];"  ";C[J],G[J]
730  NEXT J
740  PRINT
750  PRINT "MEAN SQUARED ERROR OF THE COMBINED FORECAST =",M1
760  PRINT
770  M=M-1
780  PRINT "DO YOU WISH TO REPEAT WITH ALTERNATIVE PRIOR?"
790  INPUT A$
800  IF A$="YES" THEN 100
810  END
```

SYNF/4: The Matrix Beta Outperformance Method

```
10   PRINT "INPUT THE NUMBER OF MODELS"
20   INPUT M
30   PRINT "INPUT THE OUTPERFORMANCE MATRIX"
```

```
40  MAT INPUT G[M,M]
50  PRINT "INPUT THE LENGTH OF THE TIME SERIES"
60  INPUT L
70  MAT B=ZER[M,M]
80  MAT C=ZER[M,M]
90  MAT D=ZER[M,M]
100 MAT K=ZER[M]
110 MAT M=ZER[10]
120 MAT P=ZER[M]
130 MAT B=G
140 Z=0
150 FOR I=1 TO M
160 Z=Z+B[I,I]
170 NEXT I
180 FOR I=1 TO M
190 P[I]=B[I,I]/Z
200 NEXT I
210 FOR Q=1 TO L
220 FOR J=1 TO M
230 Z=0
240 FOR K=1 TO M
250 Z=Z+B[J,K]
260 NEXT K
270 FOR K=1 TO M
280 C[J,K]=B[J,K]/Z
290 NEXT K
300 NEXT J
310 MAT D=TRN(C)
320 MAT K=D*P
330 J=1
340 IF (ABS(K[J]-P[J])<.005) THEN 370
350 MAT P=K
360 GOTO 320
370 J=J+1
380 IF(J <=M) THEN 340
390 FOR J=1 TO M+1
400 READ A[J]
410 NEXT J
420 C=0
430 FOR J=1 TO M
440 K=J+1
450 C=C+P[J]*A[K]
460 M[K]=M[K]+(A[K]-A[1])↑2
470 NEXT J
480 M[1]=M[1]+(C-A[1])↑2
490 PRINT
500 FOR J=2 TO M+1
510 K=J-1
520 PRINT "FORECAST" ;K,A[J],"SSE ";M[J]
530 NEXT J
540 PRINT
550 PRINT "COMBINED ",C,"SSE  ";M[1]
560 PRINT "ACTUAL",A[1]
570 FOR K=1 TO M
```

```
580  Z=0
590  FOR J=1 TO M
600  IF (ABS(A[K+1]-A[1])<ABS(A[J+1]-A[1])) THEN 640
610  IF (K=J) THEN 640
620  B[K,J]=B[K,J]+1
630  Z=1
640  NEXT J
650  IF (Z=1) THEN 670
660  B[K,K]=B[K,K]+1
670  NEXT K
680  NEXT Q
690  PRINT
700  PRINT "***FINAL MEAN SQUARED ERRORS***"
710  PRINT
720  FOR I=2 TO M+1
730  Z=M[I]/L
740  K=I-1
750  PRINT "MSE  ";K,Z
760  NEXT I
770  Z=M[1]/L
780  PRINT "MSE COMBINED",Z
790  PRINT
800  PRINT "POSTERIOR MATRIX BETA PARAMETERS"
810  FOR I=1 TO M
820  FOR J=1 TO M
830  PRINT B[I,J];
840  NEXT J
850  PRINT
860  PRINT
870  NEXT I
880  END
```

SYNF/5: The Inverted Wishart Procedure with 'orthodox' Uniformative Prior Facility

```
10   PRINT "INPUT THE NUMBER OF MODELS"
15   INPUT M
20   PRINT "INPUT THE MATRIX OF PRIOR INVERTED WISHART
     PARAMETERS"
25   MAT INPUT P[M,M]
30   PRINT "INPUT THE DEGREES OF FREEDOM"
35   INPUT F
40   PRINT "INPUT THE LENGTH OF THE TIME SERIES"
45   INPUT L
50   MAT W=ZER[M,M]
55   MAT X=ZER[M,M]
60   MAT Y=ZER[M,M]
65   MAT S=ZER[M]
70   S=0
75   FOR I=1 TO L
80   MAT K=ZER[M]
85   READ A
90   FOR J=1 TO M
```

```
95   READ F[J]
100  NEXT J
105  MAT Y=P
110  D=0
115  FOR K=1 TO M
120  FOR J=1 TO M
125  D=D+ABS(Y[K,J])
130  NEXT J
135  NEXT K
140  D=D/(10*M*M)
145  MAT Y=(1/D)*Y
150  D=0
155  FOR J=1 TO M
160  IF Y[J,J]>.001 THEN 175
165  K[J]=1
170  D=D+1
175  NEXT J
180  IF D >= 1 THEN 245
185  MAT X=Y
190  GOSUB 625
195  D=0
200  FOR J=1 TO M
205  FOR K=1 TO M
210  K[J]=K[J]+W[J,K]
215  NEXT K
220  D=D+K[J]
225  NEXT J
230  IF ABS(D)>.0001 THEN 245
235  MAT K=CON
240  D=M
245  C=0
250  FOR J=1 TO M
255  C=C+K[J]*F[J]/D
260  NEXT J
265  MAT E=ZER[M]
270  FOR J=1 TO M
275  E[J]=A-F[J]
280  NEXT J
285  FOR J=1 TO M
290  FOR K=1 TO M
295  P[J,K]=(P[J,K]*F+E[J]*E[K])/(F+1)
300  NEXT K
305  S[J]=S[J]+E[J]↑2/L
310  NEXT J
315  S=S+(C-A)↑2/L
320  F=F+1
325  PRINT
330  PRINT "MODEL", FORECAST","WEIGHT"
335  FOR J=1 TO M
340  W=K[J]/D
345  PRINT J,F[J],W
350  NEXT J
355  PRINT
360  PRINT "ACTUAL",A
```

```
365  PRINT "COMBINED",C
370  PRINT
375  NEXT I
380  FOR I=1 TO M
385  PRINT "MSE MODEL";I,S[I]
390  NEXT I
395  PRINT "MSE COMBINED",S
400  PRINT
405  PRINT "POSTERIOR INVERTED WISHART PARAMETER MATRIX"
410  PRINT
415  FOR I=1 TO M
420  FOR J=1 TO M
425  PRINT P[I,J]:
430  NEXT J
435  PRINT
440  PRINT
445  NEXT I
450  PRINT "DEGREES OF FREEDOM";F
455  PRINT
460  STOP
465  D=1
470  FOR K=1 TO M-1
475  GOSUB 550
480  B=X[K,K]
485  D=D*B
490  FOR J=K TO M
495  X[K,J]=X[K,J]/B
500  NEXT J
505  FOR J=K+1 TO M
510  B=X[J,K]
515  FOR Q=K TO M
520  X[J,Q]=X[J,Q]-B*X[K,Q]
525  NEXT Q
530  NEXT J
535  NEXT K
540  D=D*X[M,M]
545  RETURN
550  FOR J=K+1 TO M
555  IF ABS (X[K,K]) >= ABS(X[J,K]) THEN 590
560  FOR Q=1 TO M
565  T[Q]=X[K,Q]
570  X[K,Q]=X[J,Q]
575  X[J,Q]=T[Q]
580  NEXT Q
585  D=-D
590  NEXT J
595  RETURN
600  V=0
605  FOR K=1 TO 12
610  V=V+RND(L)-.5
615  NEXT K
620  RETURN
625  FOR X=1 TO M
630  FOR Y=1 TO M
```

```
635  FOR J=1 TO M
640  IF J=X THEN 705
645  IF J>X THEN 660
650  P=J
655  GOTO 665
660  P=J-1
665  FOR K=1 TO M
670  IF K=Y THEN 700
675  IF K>Y THEN 690
680  Q=K
685  GOTO 695
690  Q=K-1
695  X[P,Q]=Y[J,K]
700  NEXT K
705  NEXT J
710  M=M-1
715  GOSUB 465
720  M=M-1
725  W[X,Y]=D*((-1)↑(X+Y))
730  NEXT Y
735  NEXT X
740  RETURN
745  END
```

SYNF/6: The Inverted Wishart Procedure with the 'suspended' Uninformative Prior Facility

```
10   PRINT "INPUT THE NUMBER OF MODELS"
15   INPUT M
20   PRINT "INPUT THE MATRIX OF PRIOR INVERTED WISHART
     PARAMETERS"
25   MAT INPUT P[M,M]
30   PRINT "INPUT THE DEGREES OF FREEDOM"
35   INPUT F
40   PRINT "INPUT THE LENGTH OF THE TIME SERIES"
45   INPUT L
50   MAT W=ZER[M,M]
55   MAT X=ZER[M,M]
60   MAT Y=ZER[M,M]
65   MAT S+ZER[M]
70   S=0
75   FOR I=1 TO L
80   MAT K+ZER[M]
85   READ A
90   FOR J=1 TO M
95   READ F[J]
100  NEXT J
105  MAT Y=P
110  D=0
115  FOR K=1 TO M
120  FOR J=1 TO M
125  D=D+ABS(Y[K,J])
130  NEXT J
```

```
135  NEXT K
140  D=D/(10*M*M)
145  MAT Y=(1/D)*Y
150  D=0
155  FOR J=1 TO M
160  IF Y[J,J]>.001 THEN 175
165  K[J]=1
170  D=D+1
175  NEXT J
180  IF D >= 1 THEN 245
185  MAT X=Y
190  GOSUB 650
195  D=0
200  FOR J=1 TO M
205  FOR K=1 TO M
210  K[J]=K[J]+W[J,K]
215  NEXT K
220  D=D+K[J]
225  NEXT J
230  IF ABS(D)>.0001 THEN 245
235  MAT K=CON
240  D=M
245  C=0
250  FOR J=1 TO M
255  C=K+K[J]*F[J]/D
260  NEXT J
265  MAT E=ZER[M]
270  FOR J=1 TO M
275  E[J]=A-F[J]
280  NEXT J
285  FOR J=1 TO M
290  FOR K=1 TO M
295  P[J,K]=(P[J,K]*F+E[J]*E[K])/(F+1)
300  NEXT K
305  S[J]=S[J]+E[J]↑2/L
310  NEXT J
315  S=S+(C-A)↑2/L
320  F=F+1
325  IF I>1 THEN 350
330  FOR J=1 TO M
335  P[J,J]=P[J,J]*2
340  NEXT J
345  MAT P=(1/2)*P
350  PRINT
355  PRINT "MODEL", "FORECAST", "WEIGHT"
360  FOR J=1 TO M
365  W=K[J]/D
370  PRINT J,F[J],W
375  NEXT J
380  PRINT
385  PRINT "ACTUAL",A
390  PRINT "COMBINED",C
395  PRINT
400  NEXT I
```

```
405  FOR I=1 TO M
410  PRINT "MSE MODEL";I,S[I]
415  NEXT I
420  PRINT "MSE COMBINED",S
425  PRINT
430  PRINT "POSTERIOR INVERTED WISHART PARAMETER MATRIX"
435  PRINT
440  FOR I=1 TO M
445  FOR J=1 TO M
450  PRINT P[I,J];
455  NEXT J
460  PRINT
465  PRINT
470  NEXT I
475  PRINT "DEGREES OF FREEDOM";F
480  PRINT
485  STOP
490  D=1
495  FOR K=1 TO M-1
500  GOSUB 575
505  B=X[K,K]
510  D=D*B
515  FOR J=K TO M
520  X[K,J]=X[K,J]/B
525  NEXT J
530  FOR J=K+1 TO M
535  B=X[J,K]
540  FOR Q=K TO M
545  X[J,Q]=X[J,Q]-B*X[K,Q]
550  NEXT Q
555  NEXT J
560  NEXT K
565  D=D*X[M,M]
570  RETURN
575  FOR J=K+1 TO M
580  IF ABS(X[K,K]) >=ABS(X[J,K]) THEN 615
585  FOR Q=1 TO M
590  T[Q]=X[K,Q]
600  X[J,Q]=T[Q]
605  NEXT Q
610  D=-D
615  NEXT J
620  RETURN
625  V=0
630  FOR K=1 TO 12
635  V=V+RND(L)-.5
640  NEXT K
645  RETURN
650  FOR X=1 TO M
655  FOR Y=1 TO M
660  FOR J=1 TO M
665  IF J=X THEN 730
670  IF J>X THEN 685
675  P=J
```

```
680  GOTO 690
685  P=J-1
690  FOR K=1 TO M
700  IF K>Y THEN 715
705  Q=K
710  GOTO 720
715  Q=K-1
720  X[P,Q]=Y[J,K]
725  NEXT K
730  NEXT J
735  M=M-1
740  GOSUB 490
745  M=M+1
750  W[X,Y]+D*((-1)↑(X+Y))
755  NEXT Y
760  NEXT X
765  RETURN
770  END
```

APPENDIX 2: THE ESTIMATION OF PRIOR DENSITY FUNCTIONS

Introduction

An appraisal of recent work on the estimation of subjective probability distributions is developed. Particular emphasis is given to the evidence on estimation bias and the special requirements of assessing conjugate prior distributions for some of the more common statistical models, particularly those relevant to the forecasting methodologies developed in previous sections.

The line of analysis will be firstly to discuss methods for estimating the subjective probability of a single realizable proposition such as "it will rain tomorrow." This will then be extended to the estimation of a distribution over a countably infinite number of propositions. Attention will then be focused upon the assessment of certain conjugate distributions which are necessary in the analysis of important data-generating processes. Finally, a discussion of wider issues of predictive bias and some behavioral factors of implementation will follow. In the first place, however, a few points will be made on the taxonomy of assessment methods.

There has been a constant motivation by writers on this subject to classify the assessment methods as direct or indirect. Winkler (1967), Phillips and Thomas (1973) and Hampton, et al. (1973) have all adopted a mechanical rationale which refers to the way in which the probability measure is extracted from the individual. Thus a method has been called direct if the person responds by explicitly placing a number on his subjective belief in a proposition and indirect if it has to be deduced from some other mode of response such as betting behavior.

Clearly, in the strictest interpretation, the class of such direct methods is very small; excluding in fact every other mode of response apart from the explicit statement of a numerical probability. With such a strict interpretation evidently rather trivial, writers have unsurprisingly been inconsistent in their labeling of the methods.

A clarification proposed by Bunn and Thomas (1973) was to call direct those methods which set out to put a probability measure directly upon an individual's degree of belief in a proposition. In other words, they set out to measure subjective probability interpreted in the sense, for example, of Jeffreys (1961) as representing relative degrees of confidence over a set of propositions. Indirect methods were to be interpreted as attempting to indirectly impute a subjective probability measure from overt choice behavior. Subjective probability is in this case interpreted in the sense that when a rational individual, obeying for example the axioms of Savage (1954) makes a decision, he is behaving as if he attached certain probabilities to the outcomes.

In many problems of individual choice under uncertainty, the pursuit of the ideal Savage would encourage the use of indirect assessment methods. It is argued in Bunn (1976), however, that in the more macro policy decision area, it is necessary to take a forecasting approach to the formulation of subjective probability. The emphasis is then placed upon subjective probability being an estimate of some ideal logical probability which could if necessary be used in the original von Neumann and Morgenstern (1947) decision theory based upon separate empirical probabilities. Under subjective probability estimation it is legitimate to discuss questions of predictive bias which are irrelevant in Savage's behavioristic formulation. Thus, in the context of forecasting and policy analysis, attention should be restricted to the class of direct assessment methods. However, various structural assumptions can be postulated to allow indirect subjective probabilities to be considered estimates in the direct sense. De Finetti (1970) for example, imposes the "rigidity" constraint that gambling behavior should be based upon EMV grounds in order to derive a probability measure.

The proposed assessment methods are really only devices to aid the articulation of an individual's judgment in terms of a mathematical probability measure. They do not generally contribute much help to the individual in formulating his set of beliefs in the first place. Indeed, much of the research in assessment methodology is concerned with developing procedures to reduce the bias which is introduced by the actual structural form of the assessment itself.

The Estimation of the Probability of a Realizable Proposition

Since the probability of a proposition will rarely be estimated without the implicit consideration of its complement, this section will more strictly refer to the assessment of sets of realizable propositions. However, discussion will be restricted to sets of propositions of a size applicable to fans on a decision tree, but not large enough to be evaluated as "continuous" distributions.

The apparently simplest assessment procedure would be one in which the subject responded by directly articulating a set of numerical probabilities. Unfortunately psychologists (c.f. Phillips (1973)) have suggested that this is not necessarily the simplest conceptual task, particularly for a decision maker with little experience in probability. For example, different measures are obtained if the individual responds in terms of odds ratios or direct probabilities. For this reason, the use of standard devices is generally advocated as a medium of expression, some of which are described in Bunn and Thomas (1973).

It will be useful to conceptualize a subject's fundamental notions on uncertainity to be in the form of a Non-Probabilistic Chance Perception (NPCP) and that the function of the assessment procedure is one of mapping this cognitive structure into a consistent Probability Density Function (PDF). The imposition of the consistency requirement may also introduce a degree of belief formulation during the assessment procedure itself, apart from its pure operation as a transformation. Thus, standard devices attempt to furnish the individual with a physical equivalent to their NPCP from which a PDF can then easily be deduced.

A useful standard device is an urn filled with 1000 identically shaped balls. Each ball is identified with a number, from 1 to 1000. The simple experiment of drawing, blind, one ball from the urn is to be performed. Phillips and Thomas (1973) describe this method as it is often presented by a decison analyst in practice:

> To see how the standard device can be used to measure degrees of belief, we must consider two bets, one involving the event whose probability you wish to assess, and one involving the standard device.
>
> Suppose, for example, you want to determine the probability that it will rain tomorrow. Imagine that the following bet has been offered to you:
>
> Bet A { If it rains tomorrow, you win 1,000.
> If no rain tomorrow, you win nothing.
>
> The tree-diagram of Figure A1-A is a convenient representation of this bet.

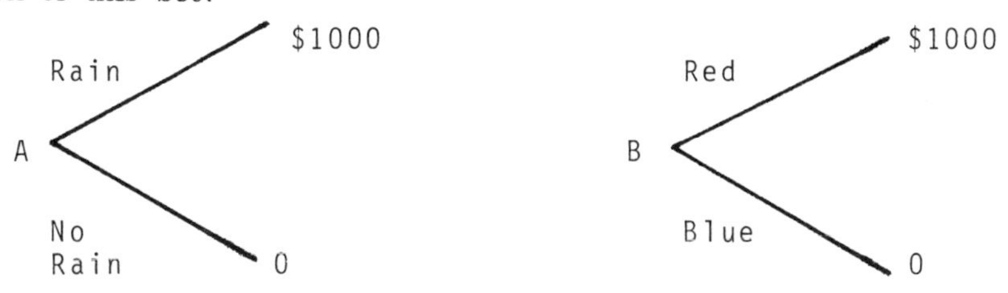

Figure A1: Tree-diagrams for the "rain-tomorrow" bet and for the reference bet.

Now imagine that balls 1 through 500 in the standard urn have been painted red while the remaining 500 balls have been colored blue. The balls are thoroughly mixed, and one is to be drawn by a blindfolded observer as tomorrow draws to a close. Now consider this bet:

Bet B { If the ball drawn is red you win 1,000.
If the ball is blue, you win nothing.

This bet is shown in Figure A1-B. We would all agree that the probabilities of drawing a red or blue ball are 0.5 respectively, and these probabilities are shown on the branches of the tree. Remember, we are trying to find out what probabilities should be shown on the branches of the tree representing the "rain tomorrow."

Consider both bets. Which do you prefer, A or B? Suppose you prefer B. Then there must be a better chance for you to win 1,000 with Bet B than with Bet A. Thus, the probability of rain tomorrow in your judgment, is clearly less than 0.5.

By changing the proportion of red balls in the urn, it is eventually possible to find a mix of red and blue balls that make you indifferent between the two bets. When this point is reached, then we are justified in assigning the same probability to the event "red ball is drawn" as we are to the event "rain tomorrow." At no time is it necessary to ask a question more complex than "Do you prefer this bet or that one, or are you indifferent between them?" Numerical measurement of an individual's subjective probability can be obtained simply by asking questions of preference.

Other standard devices have been popular. A pie diagram, or spinner, is a favorite with Stanford Research Institute. A circle is divided into two sectors and the relative sizes of the sectors can be adjusted. A spinner randomly selects one of the two sectors and hence the larger a sector the greater its chance of being chosen. The same bets as those shown in Figure A1 can be offered, but the outcomes for Bet B are determined not by drawing a ball from an urn, but by noting which sector is chosen. The relative sizes of the sectors are adjusted until the indifference point is reached; the sector sizes then represent the probabilities of the event being assessed and its complement.

The prerequisite of a standard device is that it should have easily perceived probabilistic implications, otherwise it will introduce bias. Thus Phillips and Thomas (1973) report some preliminary investigations which suggest that assessments using the urn device are 0.02 to 0.07 larger than the probabilities from the SRI spinner.

It is proposed to distinguish between task bias which is characteristic of the assessment method itself and conceptual bias which is idiosyncratic to the individual. Task bias could be caused by the standard device having misunderstood probabilistic implications or because it also structures thinking and maybe changes the fundamental beliefs of the individual in some systematic way. The more fundamental conceptual bias represents a faulty NPCP and will relate to his inability to process information and deduce the causal implications of the various inductive hypotheses. The endeavor to develop formal methods for synthesizing inductive models would be an attempt to reduce this conceptual bias, but fundamental limitations relating to an individual's perception of chance processes mitigate against this and the reduction of task bias.

Tversky and Kahnemann (1973) have recently presented an important paper dealing with a characterization of different sorts of conceptual bias. They isolate three types of systematic bias in the formulation of probabilistic judgment: representativeness, availability and adjustment.

Representativeness

Individuals apparently formulate probabilistic judgment by means of a representativeness heuristic. If x is considered highly representative of a set A, then it is given a high probability of belonging to A. However, this approach to the judgment of likelihood leads to serious bias because many of the factors important in the assessment of likelihood play no role in such judgments of similarity. One factor is the prior probability or base rate frequency. For example, given a neutral description of a person and being asked to estimate the probability of his being a lawyer or an engineer, subjects were found to answer 0.5 regardless of prior information on the relative number of lawyers and engineers in the population. Similarly, the representativeness heuristic does not take any account of sample size. Thus the manifestation of the gambler's fallacy can be ascribed to the belief that randomness is expected to be represented in even very small samples. Tversky and Kahnemann describe many fascinating cases of such bias.

Availability

Reliance on the availability heuristic introduces bias through the inadequacy of the cognitive process in conceptualizing all of the relevant information. There is a memory retrievability problem which can cause a bias such as the probability of a road accident's increasing dramatically after witnessing such an event and by much more than by just reading about it. The limitations of the memory search process cause people to judge that there are more words beginning with "r" than with "r" in the third place when in fact the converse is true. Conceptual limitations of imaginability and scenario formulation encourage subjects to believe for example that many more committees can be constructed of size 2 from 10 than of size 8. Again Tversky and Kahnemann relate many interesting examples.

Adjustment and Anchoring

In most situations it is found that indviduals formulate their general belief structure by starting from some obvious reference point and adjusting for special features. Typically, however, the adjustment is not sufficient and a bias towards these initial values is described as anchoring. Thus when subjects were asked to estimate within 5 seconds the product 8 x 7 x 6 x . . .x 1 they gave a much higher answer than those asked the product 1 x 2 x 3 . . .x 8. A much fuller consideration of anchoring will follow in the context of the fractile assessment method for a distribution in the next section.

There is now an enormous amount of published experimental work investigating how individuals deviate from Bayesian rationality in formulating their probabilistic judgment. Much of it is equivocable and the paper by Tversky and Kahnemann (1973) represents one of the few attempts at deriving an overall theory. Phillips and Edwards (1966) looked not so much at the formulation of belief but in its revision. Compared with the rational Bayesian paradigm, individuals have generally been found to be "conservative" information processors, underestimating the overall diagnosticity of observed evidence. Again, more will be said about conservatism bias in the next section in the particular context of the "imaginary results" assessment method.

One way of minimizing some of this form of bias is to ask for less precise estimates by not requiring the responses to be on a metric scale. There have been various forms of psychometric ranking methods proposed. In some cases a simple ranking of the outcomes may be sufficient or a decision analysis can be structured in such a way as to require only a sensitivity analysis of certain crucial probability parameters. However, in general, the necessary degree of precision will imply a ranking of first differences, as in Smith (1967).

The indirect estimation of probabilities from gambling preferences makes strong behavioral assumptions. Quite often it is assumed that the individual is behaving to maximize expected monetary value. Thus, returning to the gamble presented in Figure A1-A, a statement of his Certainty Equivalent (CE) for the gamble allows the probability to be imputed as CE/1000.

But to have any confidence in this as a predictive probability, it should be ascertained that an individual's utility function is in fact linear over this range. However, if trouble is to be taken in measuring the individual's utility curve in the first place, there is no reason why the payoffs should not be appropriately mapped into utility in order to obtain consistent subjective probabilities. If a von Neumann and Morgenstern utility function is derived using standard devices to articulate the probabilities presented in the artificial gambles, then the subsequent use of this function in the derivation of subjective probabilities will give valid estimates providing the individual obeys the coherence axiom. Unfortunately, this may not be the case. Phillips (1973) quotes experimental evidence from Slovic that individuals react differently in gambling situations. Some people pay more attention to the chance of winning, others to the chance of losing, while a further group seems to look mainly at the amounts of the payoffs.

Estimation of the Subjective Probability Distribution

The problem of assessing the distribution function over a countably infinite set of propositions is usually reduced to a set of discrete assessments requiring only the application of one of the methods of the previous section. Bias can, however, be introduced according to the way in which the range of propositions is split up. For the usual fractile method of assessment, Morrison (1967) designed the following questionnaire:

1. At what value of the variable, F(50), do you feel that there is a 50 percent chance that the true value of the variable will be below F(50)? This establishes the value at which CDF =

0.5.

2. Given that the true value of the variable is below F(50) at what value of the variable F(25) do you feel there is a 50 percent chance that the true value of the variable will be below this value? This establishes the value for CDF = 0.25.

3. Given that the true value is above F(50) at what value of the variable F(75) do you feel there is a 50 percent chance that the value of the variable will be below this value? This establishes the value for CDF = 0.75.

Evidently, this method of successive medial bisection will result in the set of quartiles or octiles, etc. A more recently favored fractile method is to assess the tertiles, i.e. those fractiles which split the range into three equally probable intervals. The successive extension of this to yield noniles may be attractive and there is doubtless the prospect of a hybrid fractile method generating sextiles.

This family of fractile assessment methods appears the most convenient way to estimate the distribution function over a continuous range. It is to be preferred to the direct estimation of a probability histogram over a set of pre-specified intervals on the range, which would be the basis for interpolating the PDF, since it does not involve a response in the form of a probability metric. In all the above fractile methods, the responses are in the form of equiprobable intervals.

Barclay and Peterson (1974) compared the tertile method with the PDF histogram approach and found that anchoring bias was considerably more serious in the PDF. A central interval in the PDF method only captured the "true" value 39 percent of time compared with the 75 percent for perfect calibration. In the tertile method, instead of the ideal 33 1/3 percent, the central interval captured the "true" value only 23 percent.

This tertile adjustment bias compares with the 33 percent for the 50 percent central interval in the earlier quartile experiments of Alpert and

Raiffa (1969), Tversky (1974), Winkler (1967) and Pickhardt and Wallace (1974) have all reported similar anchoring bias in the quartile method.

However, it should be recognized that all this experimental evidence is based upon artificial laboratory experimentation where the subjects will not have the same degree of motivation and personal involvement in the consequences of their probability estimates as they would in a real decision making situation. There is an obvious need for more research in this area to be undertaken in the real decision making process.

Stael von Holstein (1972) was able to use professional investment analysts in his on-going stock market forecasting and portfolio selection activity. He reported significant anchoring bias in the excessive tightness of the assessed distributions, although his choice of the PDF method may well have exacerbated this.

Winkler and Murphy (1973) were able to compare the quartile and PDF methods in the real world situation of weather forecasting. They reported that the PDF method exhibited greater anchoring bias than the quartile method for which in fact the central 50 percent interval captured the true value 45 percent of the time. They were fortunate however in having subjects with considerable experience in probabilistic forecasting; training and practice appears to have a very pronounced effect in reducing anchoring bias. Alpert and Raiffa (1969) found that after only one round in their experiment, the central 50 percent interval capture rate increased from 33 percent to 43 percent using the quartile method.

One factor which may reinforce anchoring bias is the importance generally placed upon self-consistency within the assessments. In a straightforward assessment method, it is easy for the subject to be pseudo-consistent precisely because he can perceive what he should believe in order to be consistent with his previous responses. In this way, his responses become firmly anchored from the starting point.

The author has recently published (1975) some preliminary results on a procedure which attempts to expose anchoring bias in the fractile method and thereby partially eliminate it. The method involves the derivation of an adjustment hysteresis effect where the responses are structured in such a way that it becomes difficult for the subject to exhibit pseudo-consistency. The results obtained so far are quite encouraging.

Estimation of Some Common Prior Density Functions

Univariate distributions will first be considered. It should be emphasized that this class of density functions can always be parameterized by means of deriving the individual's subjective distribution as outlined previously and then fitting the required function to it. When an interactive computer package is available, particularly if it incorporates visual display, this may well be the best available procedure. Otherwise the problem is one of estimating the parameters in their own right.

The Normal PDF

Because of symmetry, the mean can be estimated as either the mode or median and, utilizing the standard tables, the variance can easily be derived from any given set of fractiles.

The Beta PDF

The beta distribution is usually used to express prior opinion on the probability of one of the two dichotomous events in a Bernoulli process. If it is parameterized as

$$P(k) = B^{-1}(\rho + 1, \nu + 1)\, k^{\rho}(1 - k)^{\nu-\rho}$$

where $B(x,y)$ is the usual beta function and $\rho > -1$, $\nu \geq \rho$, then the mode is given by ρ/v and v is equivalent to the number of previous trials which would be required to give the same precision on an empirical basis.

Thus, the parameter v can be assessed as an Equivalent Prior Sample (EPS) which the individual feels would be the empirical equivalent of his subjective opinion. Good (1965) refers to this type of approach as one of imaginary results.

Another possibility is to elucidate how the individual's estimate of the mode would change on the basis of one more realization. This is referred to as the method of Hypothetical Future Samples (HFS). If his estimate of the prior modal probability of "success" is m^* and the posterior m^{**} is assessed after one further "failure" is envisaged, then

$$m^* = \rho/v$$

$$m^{**} = \rho/(v + 1)$$

which gives $v = m^{**}/(m^* - m^{**})$

Bayesian expectations, e^* and e^{**}, can be used instead of the model point estimates if preferred and in order that the formulae would be directly analogous, it is suggested that the more usual parameterization of the Beta distribution should be used in this case

$$P(k) = B^{-1}(\rho,v)\ k^{\rho-1}(1 - k)^{v-\rho-1}$$

with $\rho > 0$ and $v \geq \rho$.

Clearly, the methods of HFS and EPS make strong Bayesian assumptions on the way in which individuals process information. It was indicated earlier that individuals do in fact tend to be conservative processors of information. This tendency manifests itself in the assessment of too large a hypothetical sample, thus implying an excessively tight distribution similar to the anchoring bias in fractile assessment. Winkler (1967) observed this in his experiments, although the naive subjects in his groups did report an intuitive pereference for these imaginary results methods over the fractile and PDF methods.

Like most of the research in this area, our knowledge of conservatism bias is restricted to experimental behavior in the laboratory. It is quite possible that this sort of bias could be largely situational and reflect the subject's unfamiliarity with the type of data generating processes and inferential tasks with which he is confronted. For optimal behavior in these tasks, the subject must be very adept in dealing generally with the stationary Bernoulli process. The real world is characterized by non-stationarity and Phillips, Hays and Edwards (1966) have remarked that the conservatism revealed in their experiments could be caused by the subjects' believing that the data-generating process were non-stationary. Furthermore du Charme and Peterson (1968) noted an improvement in the optimality of subjects when they were dealing with Normally generated data, which they suggested was due to their greater familiarity with this type of data from the real world.

Most of the work on cascaded inference has not succeeded in revealing significant conservatism bias. Models of cascaded inference attempt to formalize the more complex inferential tasks of the real world and in fact most of probability assessments were more optimal than in the simpler experiments where conservatism bias has been most evident. Bias, if anything, has tended to be excessive rather than conservative in this general context [c.f. Youssef and Peterson (1974)].

The Inverse Gamma Distribution

This distribution is useful in the analysis of the Normal process, being the natural conjugate for the variance. It can be parameterized

$$f_{i\gamma}(y|\psi,v) = \frac{\exp(-\frac{1}{2}v\psi/y)(\frac{1}{2}v\psi/y)^{\frac{1}{2}v+1}}{\frac{1}{2}v\psi\ \Gamma\frac{1}{2}v}$$

for $y > 0$

The parameters can be given similar interpretations to those of the beta distribution. v likewise represents the size of the hypothetical prior sample and the mode is given as

$$v\psi/(v + 2)$$

Apart from using methods of imaginary results, a fractile method is possible. There is a standard result [c.f. Lavalle (1970)] connecting the fractiles of a gamma distribution to the tabulated chi-squared distribution with the same degrees of freedom. If σ_p^2 denotes the p% fractile of the assessed variance distribution, then

$$\sigma^2 p/\sigma^2 q = \chi^2(1-q)/\chi^2(1-p) \,\Big|\, v$$

e.g.

$$\sigma^2_{.75}/\sigma^2_{.5} = \chi^2_{.05}/\chi^2_{.25} \,\Big|\, v$$

and thus v can be derived from the chi-squared tables.

There are simple relations [c.f. Lavalle (1970)] between the parameters of the inverse gamma (the natural conjugate for the variance of the normal process), the inverse gamma-2 (the natural conjugate for the corresponding standard deviation) and the gamma itself which is the natural conjugate for the parameter in the Poisson process. Raiffa and Schlaifer (1961) suggest that it is probably most convenient to assess the uncertainty in terms of the standard deviation and then translate it into variance or precision.

The Dirichlet Distribution

This is the natural conjugate to the multinomial process and can be expressed:

$$f(\underset{\sim}{p}|\underset{\sim}{\rho},v) = \prod^{k} (p_i^{\rho_i - 1}) \prod^{k} \Gamma\rho_i/\Gamma v$$

with $\underset{\sim}{p}$ and $\underset{\sim}{\rho}$ k-dimensional vectors such that $\sum^{k} p_i = 1$, $\underset{\sim}{p} > 0$, $\sum^{k} \rho_i = v$ and $\underset{\sim}{\rho} > 0$. The easiest conceptual approach is probably one of hypothetical results. v represents the hypothetical sample size and ρ_i/v the expected proportions for each element. v can also be estimated according to the HFS method described for the beta distribution.

The fact that each element is marginally distributed beta means that a fractile assessment method is applicable. Each element can be marginally assessed as a beta subject to the constraint that v should be the same for each and the ρ_i sum to v.

The Inverted Wishart Distribution

The inverted Wishart distribution is the natural conjugate for the covariance matrix in the Multinormal process and can easily be seen to be a generalization of the inverse gamma density function into k dimensions.

$$f_{iw}^{(k)}(\underset{\sim}{S}|\underset{\sim}{\psi},v) = W_k(v)^{-1}|\underset{\sim}{\psi}|^{\frac{1}{2}(v+k-1)}|\underset{\sim}{S}^{-1}|^{\frac{1}{2}(v+2k)} \quad \exp(-\tfrac{1}{2}v\,tr(\underset{\sim}{S}^{-1}\underset{\sim}{\psi}))$$

with $W_k(v) = (2/v)^{k(v+k-1)/2}\pi^{k(k-1)/4}\ \Pi_{i=1}^{k}\ \Gamma\tfrac{1}{2}(v+k-i)$

defined for S positive definite and symmetric and $v>0$.

mean $(\underset{\sim}{S}) = \underset{\sim}{\psi}v/(v-2)$

mode $(\underset{\sim}{S}) = \underset{\sim}{\psi}v/(n + 2k)$

The parameter v represents the effective imaginary sample size and HFS and EPS methods are evidently feasible. It is probably easier however to consider the marginal inverse gamma distributions for the diagonal variances and then assess the matrix of intercorrelation coefficients.

Wider Issues

The main concern of this Appendix has been the examination of the structure of various subjective probability assessment procedures, their properties and characteristic biases. The wider social and political aspects of the problem have not been considered. The lack of attention to issues of subject motivation and orientation, socio-psychological factors in the decision-analyst and probability-assessor relationship, personal resistance to ambiguity and uncertainty, and training programs is not to be interpretated as a relegation of the undoubted crucial importance of these factors for successful implementation, but a recognition of the fact that many of these issues are common to the practice of operational research and not essentially structural. It should be realized, however, that the practical implementation of such highly formalized subjective techniques as these is open to even more abuse than usual. Phillips (1973) discusses many of the important factors in the assessment task which are necessary in persuading a reluctant subject to respond and in minimizing the bias which is introduced by the very presence and behavior of the analyst himself.

It is argued in Bunn (1977) that the general context of policy analysis requires the adoption of a forecasting approach to subjective probability estimation. The theory of individual choice under uncertainty is inadequate to deal with the collective policy decision of a group relying upon the probabilistic judgment of a disparate set of experts and models. There is an inevitable separation of the probability and utility assessment tasks in the orthodox decision analysis. As a consequence, however, because the probability assessors are now in a non-motivating, decision-neutral state, extra care has to be taken to minimize casual bias (recall the discussion on the generalizability of laboratory evidence upon anchoring and conservatism bias).

Where subjective probability forecasts are produced on a repetitive, team basis, a good example is the United States weather forecasters, the use of penalty functions or scoring rules (c.f. Winkler and Murphy, 1973) have been successful. Essentially, an error function is defined upon the actual result and the forecaster's probabilistic assessment such that in attempting to maximize his score, he will be increasing the accuracy of his estimation.

Another consequence of this concentration upon encoding structures is the partial elimination of consideration of the fundamental cognitive processes whereby individuals formulate their beliefs and perceptions. Very little is known about these processes and much of the research must necessarily fall within the scope of neuropsychology. The basic topic, however, is one of aggregating information and constructing sensible inferences. At normative level, the research on hierarchical inference [c.f. Peterson (1973)] and on the combination of forecasts can provide the basis for formal procedures, or at least conceptual structures, in meeting this need. The more explicit an individual's reasoning, the easier it is to identify the precise points of disagreement within a group and hence achieve a consensus. Developments along this line of explicating the "thinking algorithms" underlying an individual's subjective probability assessment are at present restricted by the lack of a suitable notational logic and representation but are evidently in the spirit of subjective probability in attempting to make statistics less, paradoxically, subjective in the sense that the subjectivity in the analysis is more clearly defined. It should be recognized, however, that control over an individual's formulation of probability judgment can never attain the level of deriving a completely "unbiased" estimate without the probability's ceasing to be subjective.

REFERENCES

Adam, E. E. (1973) Individual Item Forecasting Model Evaluation. Decision Sciences.

Aitchison, J. (1970) Choice Against Chance. Addison-Wesley.

Aitchison, J., and Sculthorpe, D. (1965) Some Problems of Statistical Prediction. Biometrika.

Aitchison, J. and Kay, J. W. (1976) Principles, Practice and Performance of Decision Making in Clinical Medicine. In White, D., and Bowen, K. (eds.) Decision Theories in Practice. Hodder and Stoughton, London.

Alpert and Raiffa (1969) A Progress Report on the Training of Probability Assessors. Unpublished Working Paper, Harvard.

Anscombe, F. J. and Aumann, R. J. (1963) A Definition of Subjective Probability. Annals of Mathematical Statistics.

Baecher, G. B. and Gros, J. G. (1975) Extrapolation of Trending Geological Bodies. RM-75-30. IIASA, Laxenburg A-2361 Austria.

Ball, R. J., and Burns, T. (1970) Some Recent Experinece in Short Term Forecasting. EFU Discussion Paper 17, London Business School.

Barclay, S., and Peterson, C. R. (1973) Two Methods for Assessing Probability Distributions. Technical Report 73.1 Decisions & Designs Inc.

Bates, J. M., and Granger, C. W. J. (1969) The Combination of Forecasts. Operational Research Quarterly.

Bell, D. E. (1975) A Decision Analysis of Objectives for a Forest Pest Problem. RR-75-34. International Institute for Applied Systems Analysis, A-2361 Austria.

Benton, W. K. (1972) Forecasting for Management. Addison Wesley.

Bierman, H., and Hausman, W. H. (1970) The Credit Granting Decision. Management Science.

Bolt, G. J. (1971) Market And Sales Forecasting. Kogan Page.

Box, G. E. P. and Jenkins, G. M. (1970) Time Series Analysis, Prediction and Control. Holden Day.

Brown, R. G. (1963) Smoothing, Forecasting and Prediction. Prentice Hall.

Brown, R. V., Kahr, A. S. and Peterson, C. (1974) Decision Analysis: An Overview. Holt, Rinehart and Winston.

Burns, T. (1974) Business Forecasting. London Business School Short Course Brochure.

Bunn, D. W. (1975) A Bayesian Approach to the Linear Combination of Forecasts. Opl. Res. Q.

Bunn, D. W. (1975) Anchoring Bias in the Assessment of Subjective Probability. Opl. Res. Q.

Bunn, D. W. (1976) Bayesian Point Estimation on the Bernoulli Parameter. Omega.

Bunn, D. W. (1977) Policy Analytic Implications for a Theory of Prediction and Decision. Policy Sciences.

Bunn, D. W., and Thomas, H. (1976) Assessing Subjective Probability in Decision Analysis. A paper presented at the 1973 NATO Conference, Luxembourg, and published in Decision Theories in Practice, edited by D. J. White and K. Bowen, Hodder and Stoughton, London.

Bunn, D. W., and Thomas, H. (1976) J. Sainsbury and the Haul of Contraband Butter, in Theorie de la Decision et Applications, FNEGE, Paris.

Bunn, D. W. and Thomas, H. (1978) Formal Methods in Policy Formulation Birkhauser Verlag Basel.

Byrnes, W. G. and Chesterton, B. K. (1973) Decision Strategies and New Ventures. Allen and Unwin.

Carnap, R. (1952) Logical Foundations of Probability.

Chambers, Mullick and Smith (1971) How to Choose the Right Forecasting Method. Harvard Business Review.

Connor, R. J. and Mosimann, J. E. (1969) Concepts of Independence for Proportions with a Generalization of the Dirichlet Distribution. Jnl. American Statistical Association.

De Finetti, B. (1970) Theory of Probability. Wiley.

De Groot, M. H. (1970) Optimal Statistical Decisions. McGraw Hill.

Dickinson, J. P. (1973) Some Statistical Results on the Combination of Forecasts. Opl. Res. Q.

Dickinson, J. P. (1975) Some Comments on the Combination of Forecasts. Opl. Res. Q.

Drake, A. W., Keeney, R. L. and Morse, P. (1972) Analysis of Public Systems. MIT Press.

Ezzati, A. (1974) Forecasting Market Shares by Markov Processes. Management Science.

Fishburn, P. C. (1964) Decision and Value Theory. Wiley.

Gaad, A. and Wold, H. (1967) The Janus Coefficient. North Holland.

Giesel, M. S. (1974) Bayesian Comparison of Simple Macro Economic Models Money, Credit and Banking.

Good, I. J. (1962) Subjective Probability as a Measure on an Unmeasurable Set. in Nagel et al. (eds) Logic Methodology and the Philosophy of Science, Stanford University.

Good, I. J. (1965) The Estimation of Probabilities. M.I.T.

Good, I. J. (1967) Bayesian Significance Test for Multinomial Distribution, Journal of the Royal Statistical Society B.

Granger, C. W. J. and Newbold, P. (1973) Some Comments on the Evaluation of Economic Forecasts. Appl. Economics.

Granger, C. W. J. and Newbold, P. (1975) Economic Forecasting: An Atheists Viewpoint, in G. A. Renton (ed) Modelling the Economy. Heinemann.

Guerts, M. D., and Ibrahim, I. B. (1975) Comparing the Box Jenkins Approach with the Exponentially Smoothed Forecasting Model Application to Hawaii Tourists. Journal of Marketing.

Halperin, M. (1961) Almost Linearly Optimum Combination of Unbiased Estimates. American Statistical Association Journal.

Huber, G. P. (1974) Multi-Attribute Utility Models. Management Science.

ICI Limited (1960's) Monographs 1, 2 & 3. Oliver and Boyd.

James, I.R. (1972) Products of Independent Beta Variables with Application to Connor and Mosimann's Generalized Dirichlet Distribution. American Statistical Association Journal.

Jeffreys, H. (1948) Theory of Probability. Oxford.

Jenkins, G. M. (1974) Contribution to the Discussion of a Paper by Newbold and Granger. J. Roy. Statist. Soc. A.

Judd, B. R., Warner Noth, D. and Pezier, J. (1974) Assessment of the Probability of Contaminating Mars. Stanford Research Institute. Menlo Park, California 94025.

Katz, J. J. (1962) The Problem of Induction. University of Chicago Press.

Keeney, R. L. (1973) A Decision Analysis with Multiple Conflicting Objectives: The Mexico City Airport. Bell Journal of Economics and Management Science.

Keeney, R. L. and Raiffa, H. (1975) Additive Value Functions in Theorie de la Decision et Applications, FNEGE, Paris.

Keeney, R. L. and Nair, K. (1976) Evaluating Potential Nuclear Power Plant Sites in the Pacific North West Using Decision Analysis. 76-1. International Institute for Applied Systems Analysis. A-2361 Austria.

Keeney, R. L. and Raiffa, H. (1976) Decisions with Multiple Objectives. Wiley.

Keeney, R. L., Wood, E. F., David, R. and Csontos, K. (1976) Evaluating Tisza River Basin Development Plans Using Multiattribute Utility Theory. CP-76-3, International Institute For Applied Systems Analysis, Schloss Laxenburg, Austria.

LaValle, I. H. (1970) An Introduction to Probability Decision and Inference. Holt, Rinehart & Winston.

Lee, T. C., Judge, G. G. and Zellner, A. (1968) Maximum Likelihood and Bayesian Estimation of Transition Probabilities. American Statistical Association Journal.

Lindley, D. V. (1965) Probability and Statistics. Cambridge University Press.

Lindley, D. V. (1971) Making Decisions. Wiley.

Makridakis and Wheelwright (1973) Forecasting for Managers. Wiley.

Martin, J. J. (1967) Bayesian Decision Problems and Markov Chains. Wiley.

Martino, J. P. (1972) Technological Forecasting for Decision Making. Elsevier.

Mincer, J. and Zarnowitz, V. (1969) The Evaluation of Economic Forecasts in Economic Forecasts and Expectations (ed. Mincer) NBER.

Moore, P. G., Thomas, H., Bunn, D. W. and Hampton, J. (1976) Case Studies in Decision Analysis. Penguin.

Morrison, D. G. (1967) Critique of Ranking Procedures in Subjective Probability Assessment. Management Science.

Murray, G. R. and Silver, E. A. (1966) A Bayesian Analysis of the Style Goods Inventory Problem. Management Science.

Neisser, V. (1967) Cognitive Psychology. Appleton Century.

Newbold, P. and Granger C. W. J. (1974) Experience with Forecasting Univariate Time Series and the Combination of Forecasts. Journal Roy. Statistical Soc. A

OECD (1972) Analytical Methods in Government Science Policy. Paris.

Ostrom, A. and Gros, J. G. (1975) Application of Decision Analysis to Pollution Control: The Rhine River Study. RM-75-6. International Institute for Applied Systems Analysis. A-2361 Austria.

Parker, G. G. C. and Segura, E. C. (1971) How To Get A Better Forecast. Harvard Business Review.

Peterson, C. (1973) Hierarchical Inference. Decisions and Designs Inc.

Phillips, L. D. (1973) The Psychology of Measuring Probability. London Business School Decision Analysis Unit. Working Paper.

Phillips, L. D. and Edwards, W. (1966) Conservatism in a Simple Probability Inference Task. J. Experimental Psychology.

Phillips, L. D., Hays and Edwards, W. (1966) Conservatism in Complex Probabilistic Inference. IEEE Human Factors in Electronics.

Phillips, L. D. and Thomas, H. (1973), Assessing Probability and Utility. Royal Economic Society Conference on Decision Analysis, Lancaster.

Pickhardt, R. L. and Wallace, J. B. (1974) A Study of the Performance of Subjective Probability Assessors. Decision Sciences.

Pitz, G. F. (1973) A Structural Theory of Uncertain Knowledge. Fourth Research Conference on Subjective Probability,Utility and Decision Making. Rome.

Popper, K. R. (1959) The Logic of Scientific Discovery. Hutchison.

Popper, K. R. (1963) Conjectures and Refutations. Routledge and Kegan Paul.

Raiffa, H. (1968) Decision Analysis. Addison-Wesley.

Raiffa, H. and Schlaifer, R. (1961) Applied Statistical Decision Theory. Harvard.

Reichenbach, H. (1949) The Theory of Probability. University of California Press.

Reid, D. J. (1969) A Comparative Study of Time Series Prediction Techniques on Economic Data. Ph.D Thesis. University Nottingham.

Rescher, N. (1973) The Coherence Theory of Truth. Oxford.

Roberts, H. V. (1965) Probabilistic Prediction. American Statistical Association Journal.

Robert and Whybark (1974) Adaptive Forecasting Techniques. Int J. Prod. Res.

Robinson, C. (1971) Business Forecasting. Nelson.

Savage, L. J. (1954) The Foundations of Statistics. Wiley.

Schlaifer, R. (1959) Probability and Statistics for Business Decisions. McGraw-Hill.

Schweitzer, T. T. (1970) Demand Consumer Expenditure in Canada, 1926-75. Economic Council of Canada. Staff Study No. 26.

Selvidge, J. (1973) Assigning Probabilities to Rare Events. Fourth Research Conference on Subjective Probability, Utility and Decision Making. Rome.

Smith L. E. (1967) Ranking Procedures and Subjective Probability Distributions. Management Science.

Stael von Holstein, C. A. J. (1970) Assessment and Evaluation of Subjective Probability Distributions. Stockholm School of Economics.

Tan, P. K. K. (1971) Forecasts and Forecasting Accuracy for Sales of Electricity. MSc Thesis, Imperial College, London.

Theil, H. (1970) Economic Forecasts and Policy. North Holland.

Thomas, H. (1972) Decision Theory and the Manager. Pitman.

Trigg, D. W. and Leach, A. G. (1967) Exponential Smoothing with Adaptive Response Rate. Opl. Res. Q.

Tversky, A. (1974) Assessing Uncertainty. Journal Roy. Statistical Soc B.

Tversky, A. and Kahnemann, D. (1973) Judgment Under Uncertainty. Fourth Research Conference on Subjective Probability, Utility and Decision Making. Rome.

Von Winterfeld, D. and Fisher, G. W. (1973) Multiattribute Utility Theory. University of Michigan Technical Report.

Wiginton, J. (1974) A Bayesian Approach to Discrimination between Models Decision Sciences.

Wills, G., et al. (1971) Technological Forecasting. Penguin.

Winkler, R. C. (1967) The Assessment of Prior Distribution in Bayesian Analysis. American Statistical Association Journal.

Winkler, R. C. (1972) Introduction to Bayesian Inference and Decision. Holt, Rinehart and Winston.

Winkler, R. and Murphy, A. (1973) Subjective Probability Forecasting in the Red World. Fourth Research Conference on Subjective Probability, Utility and Decision Making. Rome.

Wolfe, H. D. (1966) Business Forecasting Methods. Holt, Rinehart and Winston.

Wood, E. F. (1974) A Bayesian Approach to Analyzing Uncertainty Among Stochastic Models. RR-74-16. IIASA, Laxenburg A-2361 Austria.

Youssef, Z. I. and Peterson, C. (1974) Intuitive Cascaded Inferences. *Organizational Behavior and Human Performance*.

Zellner, A. (1971) *An Introduction to Bayesian Inference in Econometrics*. Wiley.

SUBJECT INDEX